CAMBRIDGE STUDIES IN LOW TEMPERATURE PHYSICS

EDITORS

Professor A.M. Goldman
Tate Laboratory of Physics, University of Minnesota

Dr P.V.E. McClintock
Department of Physics, University of Lancaster
Professor M. Springford
Department of Physics, University of Bristol

Magnetoresistance in metals

Magnetoresistance in Metals

A.B. PIPPARD, FRS

Emeritus Professor of Physics, University of Cambridge

The right of the
University of Cambridge
to print and sell
all manner of books
was granted by
Henry VIII in 1534.
The University has printed
and published continuously
since 1584.

CAMBRIDGE UNIVERSITY PRESS

CAMBRIDGE

NEW YORK   NEW ROCHELLE   MELBOURNE   SYDNEY

CAMBRIDGE UNIVERSITY PRESS
Cambridge, New York, Melbourne, Madrid, Cape Town, Singapore, São Paulo, Delhi

Cambridge University Press
The Edinburgh Building, Cambridge CB2 8RU, UK

Published in the United States of America by Cambridge University Press, New York

www.cambridge.org
Information on this title: www.cambridge.org/9780521118804

First published 1989
This digitally printed version 2009

A catalogue record for this publication is available from the British Library

Library of Congress Cataloguing in Publication data

Pippard, A.B.
Magnetoresistance in metals / A.B. Pippard.
    p. cm. – – (Cambridge studies in low temperature physics; 2)
Bibliography: p.
Includes indexes.
ISBN 0-521-32660-5
1. Free electron theory of metals.   2. Magnetoresistance.   3. Low
temperatures.   I. Title.   II. Series.
QC176.8.E4P56 1988   87-37234
530.4′1- -dc19   CIP

ISBN 978-0-521-32660-5 hardback
ISBN 978-0-521-11880-4 paperback

Contents

Preface

When a metal is subjected to a steady, uniform magnetic field, its resistivity may change in a variety of ways, depending on such factors as the choice of metal, whether the sample is monocrystalline or polycrystalline, the directions of magnetic field and current, and of course the strength of the magnetic field. In a book published in 1923 Campbell[1] cites over 600 references, of which hardly one is worth quoting nowadays. A complete and up-to-date bibliography would now run to many thousands of entries, even after the field has been ruthlessly narrowed down to only one of the many effects surveyed by Campbell, and the vast proliferation of papers on semiconductors excluded but for a few illustrative examples. I have attempted to display the leading ideas and some of the most informative experimental material as a coherent story, so that the greater part of the individual contributions may rest for ever undisturbed in the stackrooms of libraries. Nevertheless the limited number of references contain sufficient review articles to allow retrieval of the primary data if it is thought desirable.

There was a time when magnetoresistance in metals was an active research field, even if it was studied less for its own sake than as a valuable tool in the determination of Fermi surfaces. This extremely rewarding exercise has achieved, so far as elementary metals are concerned, nearly as much as is worth working for, and with its decline magnetoresistance has also lost its popularity. Yet as a phenomenon in its own right it presents a number of very interesting problems, some of them so difficult that it is no matter for surprise that they have been abandoned; there are also many loose ends which one would welcome being tidied up, if only the reward for doing so were commensurate with the devoted care the exercise would entail. It was the fear that most of what has been won might be forgotten within a few years that led me to record what I see as the principal strands of the argument. And it is the phenomenon itself, and not its applications, that I have concentrated on. This is the best excuse I can offer for the embarrassingly large number of references I find I have made to my own work and to that of Cambridge colleagues and students. For we were one of the two strongest groups in the heyday of these studies; the Moscow physicists formed the other, and if it seems that I have presented too few examples of their experimental work I must plead that it was not easy to find published curves of a quality that could be reproduced with any feeling of satisfaction.

Throughout my research career, ever since he was my Ph.D. supervisor forty years ago, I have enjoyed the friendship and help of David Shoenberg, and have drawn copiously on the ideas and factual information stored in his head and in the comprehensive collection of reprints, systematically arranged, that is a source of admiration and a model to us all. I take this opportunity to thank him for these and many other intangible benefits.

Among others who have helped I must particularly mention, with gratitude, Christopher Nex who came to my aid whenever my limited programming ability was being overtaxed, and who is responsible for the hidden inputs behind figs. 2.4, 6.5 and 6.15: Rolf Landauer for instructing me in the brand-new topics outlined at the end of chapter 6: Bob Powell for the gift of the Fermi-surface model shown in fig. 3.5: Keith Papworth for the photographs in figs. 3.5 and 3.8: and, not least, Marjory Knewstubb for her exemplary typing of the manuscript.

General remarks

Symbols

Bold characters are used for vectors ($\mathscr{E}, \mathbf{J}, \mathbf{B}$ etc.), except in conjunction with tensors, where subscript notation is more convenient ($\mathscr{E}_i, \sigma_{ij}$ etc.). Scalar quantities and the magnitude of vectors (always positive) are printed in italics.

The following symbols are used as consistently as possible throughout the book, and are only occasionally defined. If one of them is used for some other quantity it is always defined in its context.

$\mathbf{J}$ current density.

$\mathscr{E}$ electric field strength.

$\mathbf{B}$ magnetic field strength (see note below). It usually lies along the z-axis.

e electronic charge (negative).

$\boldsymbol{\alpha} = e\mathbf{B}/\hbar$ is directed opposite to $\mathbf{B}$ since e is negative.

$\mathbf{r}$ position vector of an electron.

$\mathbf{k}$ wave vector of an electron.

$\mathbf{v}$ velocity of an electron.

$\mathbf{r}_1$ and $\mathbf{r}_t$ are the components of $\mathbf{r}$ parallel to $\mathbf{B}$ and in the plane normal to $\mathbf{B}$. Usually $\mathbf{r}_1 = \mathbf{r}_z$.

$\mathbf{k}_1$ and $\mathbf{k}_t$ are the corresponding components of $\mathbf{k}$.

θ and ϕ polar and azimuthal angular coordinates defining the direction of $\mathbf{v}$ relative to $\mathbf{B}$ and (usually) $\mathscr{E}$.

E energy of an electron.

E_F Fermi energy; the subscript F attached to any quantity indicates its value at the Fermi energy; e.g. v_F = Fermi velocity at some point on the Fermi surface, defined as the locus, in k-space, $E(\mathbf{k}) = E_\mathrm{F}$.

s scalar distance, measured along a specified curve, between two points in k-space.

S scalar area of any portion of an energy surface, $E = $ constant.

dS vector element of area on an energy surface, directed normal to the surface towards points of higher energy; dS is parallel to v and has the same sign.

$\mathscr{A}_r$ projected area in real space of a closed orbit onto a plane normal to B.

$\mathscr{A}_k$ area in k-space of a closed curve formed by the intersection of a surface of constant energy with a plane normal to B.

l and τ mean free path and relaxation time, if definable, of an electron; $l/\tau = v$.

ω_c cyclotron frequency, $= 2\pi/T$, where T is time for describing a single turn of an electron orbit.

γ $= \omega_c \tau$.

σ_{ij} conductivity tensor, written as a scalar σ_0 for an isotropic (cubic) metal when $B = 0$.

ρ_{ij} resistivity tensor, written as a scalar $\rho_0 = 1/\sigma_0$ for an isotropic metal when $B = 0$.

r_0 residual resistance ratio (RRR), i.e. $\rho_{0°C}/\rho_{0K}$.

Crystallographic notation

This is mainly used to denote a direction, or the orientation of a plane, in a cubic metal.

A plane (hkl) intersects the orthogonal cartesian axes, coinciding with the cube edges, at a/h, a/k and a/l from the origin, a being a constant, usually the length of a side of the unit cell.

The direction of a line is denoted by $[hkl]$, the direction cosines with respect to the cartesian axes being $h/N, k/N$ and l/N, where $N^2 = h^2 + k^2 + l^2$.

The indices may be separated by commas to avoid ambiguity.

Only occasionally will the notation be used precisely; thus [100] or [001] usually mean any cube axis, and [111] any diagonal.

B and H

When an electron is governed by an extended wave-function, as in a metal, its interaction with a magnetic field is determined by B rather than H; that is to say, if the permeability μ is not unity the character of the orbit is determined by μH. It is preferable to forget H altogether and use B to define all field strengths. The vector potential A is correspondingly defined such that curl $\mathbf{A} = \mathbf{B}$. Except when the Shoenberg interaction is significant, and it is only mentioned in passing in this book, $\mathbf{B}$ is effectively the same inside and outside the metal sample.

Units

In much of the literature quoted, the unit of magnetic field B is the gauss. Electric fields are frequently expressed in V/cm and resistivities in Ω cm. Almost all diagrams have been relabelled where necessary to use SI units consistently.

$$1 \text{ Tesla (T)} = 10 \text{ kilogauss}$$
$$1 \,\Omega\text{m} = 10^2 \,\Omega\,\text{cm}$$

Numbering of equations and diagrams

Each equation and diagram has its chapter and running number. The chapter number is only quoted when the reference is to another chapter. Thus (32) or fig. 6 means equation (32) or fig. 6 in the same chapter; (1.32) or fig. 1.6 means equation (32) or fig. 6 in chapter 1.

References and cross-references

To avoid breaking the continuity of the text, references to the literature are made by superscript, and refer to the bibliography at the end of the book. Marginal cross-references refer the reader to relevant arguments on the page indicated.

237

1 Survey of basic principles

Although magnetoresistance is not intrinsically a low temperature phenomenon like superconductivity or superfluidity, in practice with magnetic
field strengths available in the laboratory it is hardly likely to attract
attention except in rather pure materials at low temperatures. This is not so
true in semiconductors as in metals, which are our primary concern, or in
the semi-metal bismuth whose sensitivity to modest magnetic fields, even at
room temperature, has made it a useful compact field-measuring instru-
ment since before 1886.[1] The effect is not dramatic, about 18% increase in
resistivity in a transverse field of 0.6 T, rising to a 40-fold change at 24 T as
observed by Kapitza.[2] Copper is more typical in that the same very
powerful field gave rise to a change of only 2% at room temperature.[3]
Cooling to the temperature (4 K) of liquid helium works wonders – a
reasonably pure sample of polycrystalline copper was found to increase its
resistance 14-fold in a field of 10 T,[4] and the better material available
nowadays might be expected to show a change at least 5 times larger.[5]
As for bismuth, a pure sample can have its resistance changed by a factor of
several million by applying a field of 10 T.[6]

To estimate the conditions necessary for a marked magnetoresistance,
consider a condensed Fermi gas of electrons with an approximately
spherical Fermi surface. The conductivity and resistivity are given by the
well-known formulae

$$\sigma_0 = ne^2\tau/m^* \quad \text{and} \quad \rho_0 = m^*/ne^2\tau, \tag{1.1}$$

in which n is the number of electrons per unit volume and m^* their effective
mass, not necessarily the real electron mass. The relaxation time τ describes
the time-constant for a current to die away when the sustaining electric field
is removed. When $B = 0$ the electrons travel in straight lines between
collisions, and a magnetic field can only have a significant effect on the
conductivity if it is strong enough to bend the trajectory appreciably during
a free path. The Lorentz force $e\mathbf{v} \wedge \mathbf{B}$ bends the paths into helices whose

1

axes are parallel to **B**, and the angular velocity of an electron round its particular axis is the cyclotron frequency

$$\omega_c = eB/m^*.$$ (1.2)

The mean angle turned between collisions is $\omega_c\tau$, and unless $\omega_c\tau > 1$ no great magnetoresistance effect can be expected. As we shall see presently, a large value of $\omega_c\tau$ does not guarantee magnetoresistance, but generally speaking the criterion is sound. Combining (1) and (2) we have

$$\omega_c\tau = B\sigma_0/ne,$$ (1.3)

from which the least well known quantity, m^*, has disappeared, allowing an estimate of $\omega_c\tau$. In copper, for example, with one conduction electron per atom, the atomic volume is $7.8 \times 10^{-3} \, m^3/\text{kg mole}$, so that $n = 8.5 \times 10^{28} \, m^{-3}$. At $0 \, ^\circ C$, $\sigma_0 = 6.4 \times 10^7 \, \Omega^{-1} m^{-1}$ and hence, from (3), $\omega_c\tau = 4.7 \times 10^{-3} \, B$. In the strongest field attained by Kapitza, $B = 30 \, T$, $\omega_c\tau$ was only 0.14 and it is hardly surprising the observed effect was so small. The sample used by de Launay *et al.*,[4] however, at $4 \, K$ had a conductivity 606 times higher so that, when $B = 10 \, T$, $\omega_c\tau$ was 28; the electrons are now wound into tight enough helices to execute about 4 turns between collisions, easily enough to change the conduction process considerably.

The strong effect in bismuth at room temperature arises because, containing as it does only about 10^{-5} conduction electrons per atom,[7] its resistivity is nevertheless no more than 70 times that of copper – an indication of an unusually long relaxation time. If we overlook the fact that the Fermi surface is far from spherical, so that the use of (3) can give at best a rough estimate, the outcome of its application here is that $\omega_c\tau \sim 20 \, B$, and the occurrence of magnetoresistance, even at room temperature, is made plausible. In this case conditions are especially favourable, in that bismuth is a compensated metal, with equal numbers of electrons and holes, and such metals show the strongest magnetoresistance of all. At the other extreme lie those metals which have electrons only, and no holes, and where the Fermi surface is nearly spherical. Potassium, carefully handled, closely approximates to this ideal free-electron model and has only a very weak magnetoresistance. The ideal should show none at all, a paradoxical result that will be explained in the next section.

Direct measurement of the cyclotron frequency, combined with (2), enables one to derive an effective mass (or *cyclotron mass, m_c*) to be compared with the mass m_e of a free electron. In general a metal is not characterized by a single value of m_c/m_e, different orbits presenting different values. In bismuth, when **B** lies along the bisectrix, it can be as low as 8×10^{-3}, but there are other orbits for which it is 0.21.[7] In the noble and alkali metals it is not far from unity for the majority of orbits; but in transition and rare earth metals the electrons in the lower bands frequently have m_c/m_e considerably in excess of unity. And the mass is still further enhanced if the electron carries a clothing of virtual phonons with it, being as high as 90 in U Pt$_3$.[8]

The free-electron gas shows no magnetoresistance

In the free-electron model the electrons are supposed to move independently, obeying Newton's laws of motion and possessing a mass m (or m^* in the quasi-free model) which is isotropic. Extension to anisotropic mass, with an acceleration law $m_{ij}\dot{v}_j = F_i$, is easy, but we shall not concern ourselves at the moment with this rather artificial concept. The argument that follows does not depend greatly on the electrons forming a degenerate Fermi gas as in most metals at ordinary or low temperatures – it is equally good for the rarefied Boltzmann gas typical of pure semiconductors. Whatever the distribution of velocities, each electron is accelerated by an electric field according to the equation

$$m^*\dot{v} = e\mathscr{E}. \tag{1.4}$$

If the electrons are not deflected, and their motion randomized, by collisions the current density rises steadily,

$$\mathbf{J} = \sum e\dot{v} = ne^2\mathscr{E}/m^*, \tag{1.5}$$

the summation being taken over all electrons in unit volume.

Now let us introduce collisions by supposing that $\mathbf{J}$, after $\mathscr{E}$ is removed, would decay exponentially. This is a good approximation for an isotropic degenerate gas, and fairly good for a Boltzmann gas, and it leads us to extend (5) to read

$$\dot{\mathbf{J}} = ne^2\mathscr{E}/m^* - \mathbf{J}/\tau, \tag{1.6}$$

τ being assumed to take the same value for all directions of $\mathbf{J}$. In the steady state $\dot{\mathbf{J}} = 0$, and $\mathbf{J} = \sigma_0\mathscr{E}$, with σ_0 having the same form as in (1). It is helpful for the next stage of the argument to consider the momentum balance in the electron gas. The momentum density $\mathbf{P} = \sum m^*\mathbf{v} = m^*\mathbf{J}/e$, and (6) may be rewritten

$$\dot{\mathbf{P}} = ne\mathscr{E} - \mathbf{P}/\tau. \tag{1.7}$$

The momentum density is subject to change from two force-densities, the electrical force-density $ne\mathscr{E}$ and the collisional $-\mathbf{P}/\tau$ which is the Newtonian reaction of the collision centres in the lattice when struck by

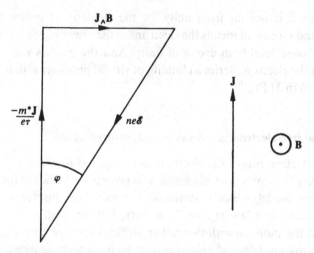

Figure 1.1 Forces acting on unit volume of a quasi-free electron gas carrying current **J** in transverse field **B**.

electrons. In the steady state, of course, they balance.[†] Now let **B** be applied transverse to **J**, exerting a Lorentz force $e\mathbf{v} \wedge \mathbf{B}$ on each electron and a force-density $\sum e\mathbf{v} \wedge \mathbf{B}$, i.e. $\mathbf{J} \wedge \mathbf{B}$, on the electron assembly as a whole. Adding this into (7) we see that in the steady state,

$$0 = ne\mathscr{E} - \mathbf{P}/\tau + \mathbf{J} \wedge \mathbf{B} = ne\mathscr{E} - m^*\mathbf{J}/e\tau + \mathbf{J} \wedge \mathbf{B}. \qquad (1.8)$$

In representing this result as a vector diagram (fig. 1) it has been assumed that neither m^* nor τ are affected by B – almost always a safe assumption. One sees immediately that the component of $\mathscr{E}$ parallel to J is unchanged, but a transverse component arises to cancel the Lorentz force. This is the Hall[(9)] field, of magnitude JB/ne. In an experiment where a current is passed along a wire or strip the direction of J is predetermined, and $\mathscr{E}$ adjusts itself, with the aid of space charges or surface charges if necessary, to satisfy (8). Then potential contacts can be placed to respond either to the parallel component, giving the resistance (unaffected by **B** in this case) or the Hall field. The angle φ between J and $\mathscr{E}$, the Hall angle, may also be measured in this way. The diagram shows that

$$\tan \varphi = eB\tau/m^* = \omega_c\tau \quad \text{from (2).} \qquad (1.9)$$

Alternatively, and especially in semiconductor physics, $e\tau/m^*$ may be

[†] The correctness of this treatment depends on the convection current of momentum being non-divergent. If this is not the case, as in an example treated in chapter 6, an extra term must be added to the momentum balance. Here, however, spatial uniformity makes it unnecessary.

written as μ, the mobility, being the drift velocity acquired by an electron in an electric field of unit strength; and $\tan \varphi = \mu B$.

The magnetoconductivity and magnetoresistivity tensors

When the direction and magnitude of **J** are fixed by the experimental procedure, and $\mathscr{E}$ adjusts itself accordingly, what is measured is the resistivity tensor, $\rho_{ij} J_j$, or a component or combination of components;

$$\mathscr{E}_i = \rho_{ij} J_j. \tag{1.10}$$

For theoretical purposes, however, it is usually more convenient to imagine $\mathscr{E}$ applied in some chosen direction and to calculate the resulting **J**, related to it by the conductivity tensor,

$$J_i = \sigma_{ij} \mathscr{E}_j. \tag{1.11}$$

The relation between σ_{ij} and its inverse ρ_{ij} is

$$\rho_{ii} = (\sigma_{jj}\sigma_{kk} - \sigma_{jk}\sigma_{kj})/\Delta(\sigma) \quad ; \quad \rho_{ij} = (\sigma_{ik}\sigma_{kj} - \sigma_{ij}\sigma_{kk})/\Delta(\sigma) \tag{1.12}$$

in which $\Delta(\sigma)$ is the determinant whose elements are the elements of σ_{ij}. Similarly,

$$\sigma_{ii} = (\rho_{jj}\rho_{kk} - \rho_{jk}\rho_{kj})/\Delta(\rho) \quad ; \quad \sigma_{ij} = (\rho_{ik}\rho_{kj} - \rho_{ij}\rho_{kk})/\Delta(\rho). \tag{1.13}$$

In the absence of a magnetic field, σ_{ij} and ρ_{ij} are symmetrical, i.e. $\sigma_{ij} = \sigma_{ji}$, and this implies that orthogonal axes can be found, with reference to which they are diagonal. In cubic, tetragonal or orthorhombic crystals the edges of the unit cell are automatically axes with this property. In cubic crystals $\sigma_{ii} = \sigma_{jj} = \sigma_{kk}$, and both ρ_{ij} and σ_{ij} are isotropic – for all directions of $\mathscr{E}$, $\mathbf{J} = \sigma\mathscr{E}$ and is parallel to $\mathscr{E}$. In tetragonal crystals two diagonal elements are the same, and in orthorhombic crystals all are different. Then with reference to different axes the off-diagonal elements do not in general vanish. Hexagonal crystals behave like tetragonal in that σ_{zz}, along the hexad axis, is different from the other two which are identical.

When B is present σ_{ij} and ρ_{ij} are not in general symmetrical, and require all nine elements of each for a complete specification. Quite frequently, however, the coupling between longitudinal and transverse effects is small enough to be neglected; that is to say, $\mathscr{E}_z$ applied parallel to **B** produces only J_z with negligible J_x and J_y, while if $\mathscr{E}$ lies transverse, in the plane normal to **B**, so also does **J**, though not necessarily parallel to $\mathscr{E}$. This verbal description is the same as putting $\sigma_{xz}, \sigma_{zx}, \sigma_{yz}, \sigma_{zy}$ all equal to zero; and similarly for the same components of ρ_{ij}. Then (12) may be written

$$\left. \begin{array}{lll} \rho_{xx} = \sigma_{yy}/\Delta'(\sigma), & \rho_{yy} = \sigma_{xx}/\Delta'(\sigma), & \rho_{zz} = 1/\sigma_{zz}, \\ \rho_{xy} = -\sigma_{xy}/\Delta'(\sigma), & \rho_{yx} = -\sigma_{yx}/\Delta'(\sigma), & \end{array} \right\} \tag{1.14}$$

6   *Survey of basic principles*

where
$$\Delta'(\sigma) = \sigma_{xx}\sigma_{yy} - \sigma_{xy}\sigma_{yx};$$
and, of course, similarly for inverting ρ_{ij} into σ_{ij}.

The behaviour represented in fig. 1 may be written, by use of (9),
$$\rho_{xx} = \rho_{yy} = 1/\sigma_0 \quad \text{and} \quad \rho_{xy} = -\rho_{yx} = \omega_c\tau/\sigma_0. \tag{1.15}$$
Also, since the conductivity along the direction of B is unaffected, $\rho_{zz} = 1/\sigma_0$, and there is no longitudinal–transverse coupling. Hence (14) applies here, and
$$\sigma_{xx} = \sigma_{yy} = \sigma_0/(1 + \gamma^2) \quad \text{and} \quad \sigma_{xy} = -\sigma_{yx} = -\gamma\sigma_0/(1 + \gamma^2) \tag{1.16}$$
where $\gamma = \omega_c\tau$. Alternatively, we may use (9) to write
$$\sigma_{xx} = \sigma_{yy} = \sigma_0 \cos^2\varphi \quad \text{and} \quad \sigma_{xy} = -\sigma_{yx} = -\tfrac{1}{2}\sigma_0 \sin 2\varphi. \tag{1.17}$$

When the direction of $\mathscr{E}$ is predetermined, rather than J as in fig. 1, the corresponding diagram is that in fig. 2, the semicircle on which the end of J lies being drawn with diameter $\sigma_0\mathscr{E}$. With increase of B the magnitude of J decreases steadily according to (16), but the resistivity ρ_{xx} remains constant since the component of $\mathscr{E}$ parallel to J decreases at the same rate.

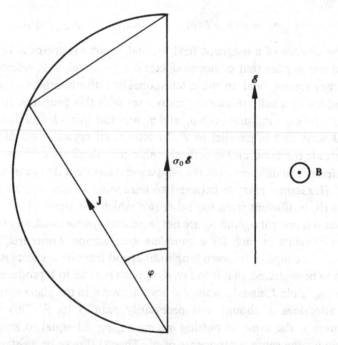

Figure 1.2 Relationship between J and $\mathscr{E}$ for a quasi-free electron gas in transverse field B.

In (15)–(17) the signs of σ_{xy} etc. depend on the sign of $\omega_c \tau$. In the diagrams in this book where cartesian coordinates are used or implied, z and $\mathbf{B}$ will point, wherever convenient, out of the page. A free electron moving in the plane of the page describes an anticlockwise orbit. The Hall coefficient R_H, defined as having magnitude $|\rho_{xy}/B_z|$, is conventionally taken to be negative for free electrons and for any conductor in which φ has the same sign as for a free-electron metal. Only rarely shall we be interested in the sign of R_H and no trouble will be taken to maintain a consistent convention – the sign can usually be adjusted by inspection at the end of any calculation.

In real metals, if $\mathbf{B}$ happens to lie along an axis of threefold or fourfold symmetry, σ_{ij} automatically possesses the properties represented in (14), being invariant with respect to rotation of the axes around $\mathbf{B}$. For rotation through ϕ makes the following transformations:

$$\sigma_{xx} \to \sigma_{xx} \cos^2 \phi - (\sigma_{xy} + \sigma_{yx}) \cos \phi \sin \phi + \sigma_{yy} \sin^2 \phi$$

and

$$\sigma_{xy} \to \sigma_{xy} \cos^2 \phi + (\sigma_{xx} - \sigma_{yy}) \cos \phi \sin \phi - \sigma_{yx} \sin^2 \phi.$$

If the axis has threefold symmetry, putting $\phi = 120°$ must leave σ_{xx} and σ_{xy} unchanged, and it is easily seen that this requires $\sigma_{yy} = \sigma_{xx}$ and $\sigma_{xy} = -\sigma_{yx}$; similarly for $\phi = 90°$, if there is fourfold symmetry about $\mathbf{B}$. Also there is no longitudinal–transverse coupling and σ_{ij} is specified by only three components which we shall frequently write as $\sigma_1 = \sigma_{xx} = \sigma_{yy}, \sigma_2 = \sigma_{xy} = -\sigma_{yx}$ and $\sigma_3 = \sigma_{zz}$; that is to say,

$$\sigma_{ij} = \begin{bmatrix} \sigma_1 & \sigma_2 & 0 \\ -\sigma_2 & \sigma_1 & 0 \\ 0 & 0 & \sigma_3 \end{bmatrix} \tag{1.18}$$

and correspondingly for ρ_{ij}.

It is sometimes convenient to treat the plane normal to $\mathbf{B}$ as a complex plane in which $\mathscr{E}$ and $\mathbf{J}$ are represented by complex numbers. They are linearly related by complex σ or $\rho, \sigma = \sigma_1 - i\sigma_2$ and $\rho = \rho_1 - i\rho_2$. The negative sign arises because $J_y = \sigma_{yx}\mathscr{E}_x = \text{Im}[\sigma\mathscr{E}]$ when $\mathscr{E}$ lies along the real axis. The inversion of σ into ρ is now a simple matter of complex algebra;

$$\rho = 1/\sigma = (\sigma_1 + i\sigma_2)/(\sigma_1^2 + \sigma_2^2) \tag{1.19}$$

so that $\rho_1 = \sigma_1/(\sigma_1^2 + \sigma_2^2)$ and $\rho_2 = -\sigma_2/(\sigma_1^2 + \sigma_2^2)$ in agreement with (14). Corresponding to (15) and (16) we have, for a free-electron metal (for which $\gamma > 0$),

$$\rho = \rho_0(1 - i\gamma) \quad \text{and} \quad \sigma = \sigma_0/(1 - i\gamma). \tag{1.20}$$

Magnetoresistance in real metals

Let us now look at some characteristic examples of magnetoresistance in real metals to see how far they depart from the free-electron model for which no change is predicted. In due course we shall be concerned to understand them in detail but for the moment they serve to illustrate the variety of forms which makes the subject interesting. The curves have been redrawn from published data, experimental points being omitted since the shapes are not in dispute. When it is reasonable to do so values of B are quoted directly and also converted by use of (3) into $\omega_c\tau$ (some authors disconcertingly present values of $\omega_c\tau$ for metals which do not approximate to the free-electron model, without making clear how they are calculated). We shall see that different metals may present widely different changes of resistance at the same $\omega_c\tau$.

It is conventional to express the degree of purity of the sample by its residual resistance ratio (RRR in many papers, but r_0 here), being the ratio of its resistance at 0 °C to its residual resistance, for which the resistance at 4 K is usually an adequate measure; and a large fraction of the measurements to be discussed were taken at this temperature in a bath of liquid helium boiling at atmospheric pressure. The magnitude of the magneto-resistance is usually expressed by $\Delta R(B)/R_0$, R_0 being the sample resistance in zero magnetic field, $R(B)$ the resistance in field $\mathbf{B}$, and $\Delta R(B) = R(B) - R_0$. In all the examples presented here the field was transverse to the current.

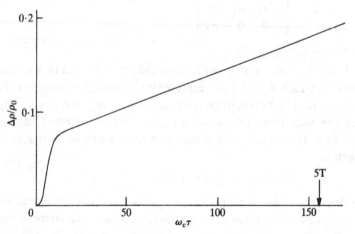

Figure 1.3 Transverse magnetoresistance at 4 K of potassium (Simpson[10]).

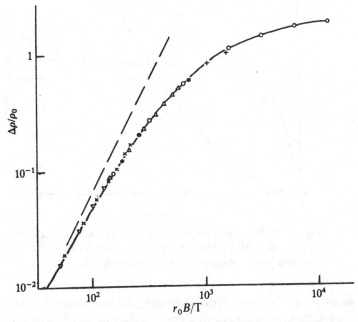

Figure 1.4 Kohler plot for transverse magnetoresistance of polycrystalline indium (Olsen[11]). The broken line shows the slope for quadratic dependence.

The metal that most closely resembles the free-electron model is potassium, and fig. 3 shows how small the magnetoresistance may be in a good sample. The linear rise, extending to the highest values of $\omega_c \tau$ attained, is not explained by any simple theory, but this is only one of the mysteries presented by potassium, as will be discussed in chapter 5. All the same, the very fact that one can achieve a value of $\Delta R/R_0$ that is 1000 times less than $\omega_c \tau$ should be seen as confirmation of the paradoxical vanishing of magnetoresistance in the ideal metal.

186

The linear variation never extends back to zero field; since field reversal normally leaves the resistance unchanged one must expect $\Delta R/R_0$ to vary as B^2 in weak fields at least, and this is confirmed fairly well, as fig. 4 shows. The quadratic range may be very short, however, as in curve (*a*) of fig. 5, where it is hardly discernible, or may continue up to high values of $\omega_c \tau$, as in curve (*b*). Both curves relate to single crystals of tin, with **B** in different directions, and it should be noted that while (*a*) saturates with $\Delta R/R_0 \sim 6$, (*b*) is still rising sharply at 750. The fact that $\Delta R/R_0$ in curve (*b*) is not far from $(\omega_c \tau)^2$, itself a rather dubiously defined quantity, is not a coincidence,

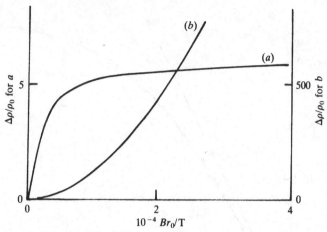

Figure 1.5 Transverse magnetoresistance at 4 K of a single crystal of tin ($r_0 = 11\,000$) with **J** along the c-axis and two different orientations of **B** in the basal plane (Alekseevskii and Gaidukov[12]).

30   as we shall see. The same holds in bismuth, in a pure sample of which a field of 1 mT sufficed to bring $\omega_c\tau$ to unity.[6] Values of $\Delta R/R_0$ well in excess of 10^6 are therefore to be expected when B approaches 10 T.

It will be observed in both fig. 4 and fig. 5(a) that a slow linear rise, like that in potassium, seems to supervene at high field strengths, marring what might have been complete saturation. This should not be confused with a dominant linear variation such as Kapitza[3] surmised might be the norm. Limited as he was, at the time of his high-field experiments, to temperatures no lower than that of liquid nitrogen, he rarely achieved values of $\omega_c\tau$ above 5. As the initial portion of fig. 5(a) exemplifies, there may well be a long, nearly linear stretch before saturation begins, and it is only with a few metals such as polycrystalline copper (fig. 6) that it continues up to high values of $\omega_c\tau$. Later work, in fact, has provided no support for what was sometimes referred to as Kapitza's law, except for certain polycrystalline

184   samples which demand special theoretical treatment.

There are many metals that can be treated adequately by imagining the conduction electrons to behave much like classical particles. The effect of **B** is solely to cause them to be deflected into orbits which are, however, frequently more complicated than the helices described by free electrons. In these, so long as the resistivity is governed by scattering processes which depend on the velocity of an electron and its direction of motion, but not on **B** or on position in the sample, the magnetoresistance is positive as in the examples so far presented. In ferromagnets, especially when they are close

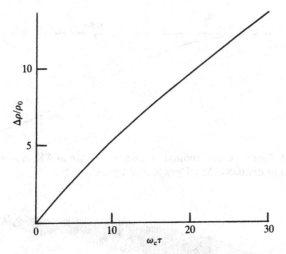

Figure 1.6 Transverse magnetoresistance at 4 K of polycrystalline copper (de Launay *et al.*[4]).

to their Curie point, the polarization of the electrons and hence their scattering can be affected by **B** and lead to negative magnetoresistance, as in fig. 7. Spatial variation of scattering plays a part in samples whose dimensions are comparable to the electronic free path, for the winding up of the orbits may reduce the importance of surface scattering and produce

200

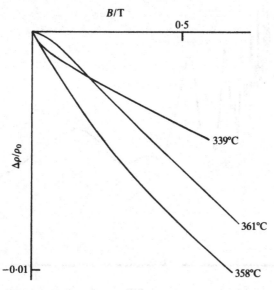

Figure 1.7 Negative transverse magnetoresistance of nickel near its Curie temperature (357 °C) (Potter[13]).

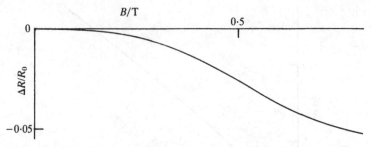

Figure 1.8 Negative longitudinal magnetoresistance at 4 K of a sodium wire, $20\,\mu$ in diameter (MacDonald and Sarginson[14]).

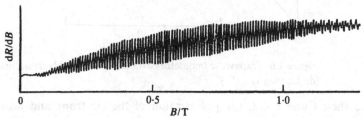

Figure 1.9 Differential transverse magnetoresistance at 1.3 K of a flat single-crystal plate of gallium, showing size-effect oscillations (Munarin and Marcus[15]).

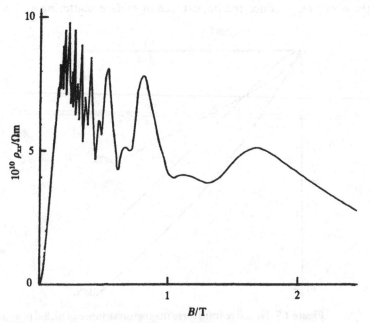

Figure 1.10 Transverse magnetoresistance at 1.6 K of a single crystal of zinc, with **B** along the hexad axis (Stark[16]).

negative magnetoresistance (fig. 8). Sometimes, however, the resistance change shows an oscillatory component, which may not appear striking when observed directly but which can be distinguished from the smooth variation by recording dR/dB instead of R. The oscillatory component thus revealed in a flat plate of gallium is shown in fig. 9, where it can be seen that the period of oscillation is constant and unaffected by the magnitude of **B**. This is characteristic of size-effect oscillations.

On the other hand, oscillations of resistance having their origin in quantization of the electron orbits show a constant period only when plotted against $1/B$. In a direct plot against B, as in fig. 10, the period lengthens markedly towards higher field strengths.

This concludes the preliminary exposure of typical magnetoresistance behaviour which it is the business of this book to explain in as much detail as possible. The rest of the chapter is concerned with the general theoretical concepts which form the basis of the variety of special treatments required.

Dynamics of an electron

In a Fermi gas of electrons at 0 K, all states up to the Fermi energy are filled and those above are empty. On raising the temperature, thermal excitation affects only those within a few $k_B T$ of the Fermi surface, a very small fraction at the temperature of most measurements. Only these few can suffer collisions, the rest having no empty state available into which they could be scattered. We visualise the conduction process as involving the bodily shift, by the action of $\mathscr{E}$, of the k-vectors of all filled states. Most states which were initially filled remain so after the application of $\mathscr{E}$ and therefore play no part in forming a current. The changes that matter are that a few empty states near the Fermi surface become filled and a few on the opposite side, that initially were filled, become empty. It is these that carry the current, and the scattering of these electrons round the Fermi surface that destroys it.

It is clear from the high conductivity of pure metals (when **B** = 0) at low temperatures that the free path, l, of electrons near the Fermi surface must be long; let us make an estimate for a free-electron metal. The relaxation time, τ, in (1) may be replaced by l/v_F, v_F being the Fermi velocity, and $m^* v_F$ is the Fermi momentum, $\hbar k_F$. Also the volume of the Fermi sphere is $\frac{4}{3}\pi k_F^3$ and this volume is packed with $1/4\pi^3$ states per unit volume of metal, so that $n = k_F^3/3\pi^2$; hence,

$$\rho l = (3\pi^2/n^2)^{1/3}\hbar/e^2. \tag{1.21}$$

In potassium, $n = 13.94 \times 10^{27}\,\mathrm{m}^{-3}$ and $\rho l = 2.19 \times 10^{-15}\,\Omega\mathrm{m}^2$. A very good sample ($r_0 = 10^4$) has $\rho \sim 6 \times 10^{-12}\,\Omega\text{m}$, so that $l \sim 1/3\,\text{mm}$. In some metals free paths of at least several mm have been attained.

Collisions between electrons do not affect the electrical conductivity of a free-electron metal, since conservation of momentum in the collision implies conservation of electric current. When different electrons have a different ratio of quasi-momentum $\hbar\mathbf{k}$ to velocity $\mathbf{v}$, however, conservation of the former in a collision implies non-conservation of the latter, and hence of the electric current. There may then be a detectable component in the resistivity attributable to electron–electron collisions, but it is not nearly so important as the effect of such collisions on the thermal current (remember the central importance of intermolecular collisions in determining the thermal conductivity of a gas). Observations[17] that the Wiedemann–Franz law is well obeyed by many metals at low temperatures provide evidence for the small importance of electron–electron collisions. Although the interactions of electrons with one another and with the ionic lattice are responsible for the often complicated band-structure, in the end it turns out that we may continue to imagine the electrons as independent particles, but not obeying the simple Newtonian laws of dynamics with constant mass. For many purposes, indeed until explicitly quantum-mechanical processes need to be considered, we may treat the electrons as classical, moving in straight lines between collisions when $\mathbf{B} = 0$, or in curved trajectories when $\mathbf{B} \neq 0$. But it helps to derive their equations of motion if we keep in mind their wavelike character, ascribing to each electron a Bloch wave-vector $\mathbf{k}$.

The energy $E(\mathbf{k})$ may be a complicated function of $\mathbf{k}$, but whatever its form an electron of wave-vector $\mathbf{k}$ moves with velocity $\mathbf{v}$ such that

$$\mathbf{v} = \hbar^{-1}\text{grad}_k E = \hbar^{-1}(\partial E/\partial k_x, \partial E/\partial k_y, \partial E/\partial k_z). \tag{1.22}$$

It is convenient to represent the form of $E(\mathbf{k})$ by drawing surfaces in $\mathbf{k}$-space on which E is constant (for free electrons the surfaces are spheres of radius $(2mE)^{1/2}/\hbar$). Since one can move in any direction along such a surface without changing E, $\text{grad}_k E$ must be a vector directed normal to the surface, pointing towards regions of higher energy. If we write this as $\partial E/\partial k_n$ it follows that the separation of neighbouring surfaces, δE apart in energy, is

$$\delta k_n = \delta E/\hbar v. \tag{1.23}$$

The rules governing the dynamical behaviour of an electron, considered as a classical particle, are tabulated below. The lattice is assumed perfect, so that there is no scattering; this is taken into account later.

1. The electron carries normal electronic charge e (unlike, for example, an excitation in a superconductor).

2. It responds to forces as if $\hbar\mathbf{k}$ were its momentum. Under the influence of $\mathscr{E}$ and $\mathbf{B}$,

$$\hbar\dot{\mathbf{k}} = \mathbf{F} = e(\mathscr{E} + \mathbf{v} \wedge \mathbf{B}). \qquad (1.24)$$

3. If $\mathscr{E} = 0$ and the electron moves under the influence of $\mathbf{B}$ alone, $\mathbf{k}$ can only move along a constant-energy surface since $\mathbf{v}$ is normal to the surface. And since $\dot{\mathbf{k}}$ is also normal to $\mathbf{B}$, $\mathbf{k}$ moves along the line of intersection of the energy surface by a plane normal to $\mathbf{B}$. If $\mathbf{k}_t$ is the component of $\mathbf{k}$, and $\mathbf{v}_t$ the component of $\mathbf{v}$, in this plane, (24) may be written

$$\hbar\dot{\mathbf{k}}_t = e\mathbf{v}_t \wedge \mathbf{B}, \quad \text{or} \quad \dot{s} = ev_t B/\hbar = \alpha v_t, \qquad (1.25)$$

s being measured around the line of intersection which will in future be referred to as the k-orbit. And of course the component of $\mathbf{k}$ along $\mathbf{B}$ does not change, $\dot{k}_z = 0$.

4. Corresponding to the changes of $\mathbf{k}_t$, the electron describes an orbit in real space (the r-orbit). Consider first the component $\mathbf{r}_t$ of $\mathbf{r}$ in a plane normal to $\mathbf{B}$; $\dot{\mathbf{r}}_t = \mathbf{v}_t$. Then (25) may be written

$$\hbar\dot{\mathbf{k}}_t = -e\mathbf{B} \wedge \dot{\mathbf{r}}_t,$$

or

$$\mathbf{k}_t = -\boldsymbol{\alpha} \wedge (\mathbf{r}_t + \mathbf{r}_0) \qquad (1.26)$$

in which $\mathbf{r}_0$ is an arbitrary constant of integration. This shows that the r-orbit is related to the k-orbit by a scaling factor α^{-1} and rotation in the transverse plane through a right angle. Fig. 11

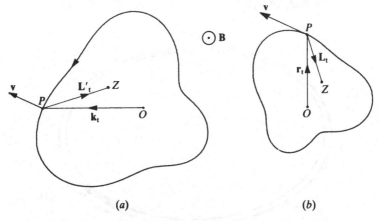

(a)   (b)

Figure 1.11 The relationship between orbits in (a) k-space and (b) r-space, as given by (26); the points P define $\mathbf{k}_t$ and $\mathbf{r}_t$, relative to O, for a typical electron. Z, $\mathbf{L}_t$ and $\mathbf{L}'_t$ are relevant to a later use of the same diagram.

illustrates this simple but fundamental result whose importance was independently realized by Onsager and I.M. Lifshitz.[18]

5. The implication of §4 is that once the shape of an E-surface is known the shapes of all possible r-orbits for an electron of this energy are also known. In addition we may determine the rate at which the orbit is executed and the accompanying motion along the direction of $\mathbf{B}$; these will be considered separately in §§6 and 7.

6. For motion in the transverse plane, note that if $\mathbf{v}$ and the normal separation $\delta\mathbf{k}_n$ in (23) make an angle θ with $\mathbf{B}$, the separation of two energy surfaces, measured in the transverse plane, $\delta k_t = \delta k_n \operatorname{cosec} \theta$, while $v_t = v \sin \theta$ (see fig. 12). Thus (23) takes the same form when applied to the transverse plane,

$$\delta k_t = \delta E/\hbar v_t \qquad (1.27)$$

which may be combined with (25) to yield

$$(\delta \dot{\mathscr{A}}_k) = \dot{s}\delta k_t = \alpha\delta E/\hbar. \qquad (1.28)$$

Here $\delta\mathscr{A}_k$ is the area of the annulus in fig. 12 between the two sections of energy surfaces δE apart, and $(\delta\dot{\mathscr{A}}_k)$ is written for the constant rate at which this area may be imagined being swept out as the normal δk_t moves round the k-orbit. The element $\delta s \cdot \delta k_t$ changes shape as it sweeps round the orbit, but not its area – an example of Liouville's theorem. The cyclotron frequency ω_c follows from (28), for if T is the time required to sweep out the complete annulus,

$$T = \delta\mathscr{A}_k/(\delta\dot{\mathscr{A}}_k)$$

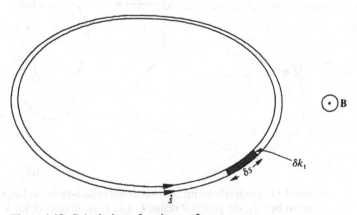

Figure 1.12 Calculation of cyclotron frequency, ω_c.

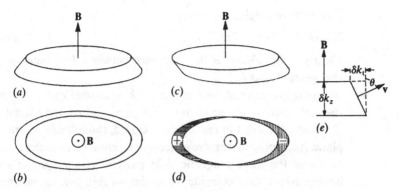

Figure 1.13 Calculation of orbit pitch (Z in (30)).

and

$$\omega_c = 2\pi/T = (2\pi\alpha/\hbar)/(\partial\mathscr{A}_k/\partial E)_{k_z}. \tag{1.29}$$

In writing this, small differences have been replaced by infinitesimals.

7. For motion parallel to **B** we need to know how the shape of an energy surface changes with k_z. Fig. 13 shows two slices, (a) and (c), δk_z thick, cut normal to **B**, and the same projected onto the transverse plane in (b) and (d). In (a) and (b) the longitudinal component of velocity v_z has the same sign everywhere, and as the electron describes its r-orbit it always moves in the same direction parallel to **B**, though not in general at constant speed. The complete trajectory is a monotonic 'helix', exactly repeated at each turn apart from its longitudinal displacement. In (c) and (d), on the other hand, the motion along **B** alternates in sign, the shaded regions of (d) marking those parts where the sign of v_z is constant.

To find v_z we adopt a procedure similar to §6 except that the annulus is now typically that in (b) or (d), the projected area $\delta\mathscr{A}'_k$ between two slices of the same energy surface. In fig. 13(e) the plane section shown contains both **B** and **v**, and δk_t is the width of the annulus; $\delta k_t = \delta k_z \cot\theta$. According to (25) $\dot{s} = \alpha v \sin\theta$, and hence

$$(\delta\cdot\mathscr{A}'_k) = \dot{s}\delta k_t = \alpha v \delta k_z \cos\theta = \alpha\dot{z}\delta k_z.$$

The displacement z is proportional to the annular area swept out and the pitch, Z, of the helix, the displacement in one cycle, is

$$Z = \alpha^{-1}(\partial A_k/\partial k_z)_E. \tag{1.30}$$

Alternatively, though more for elegance than utility, we may use

(29) to write (neglecting signs)

$$\bar{v}_z = \omega_c Z/2\pi = \hbar^{-1}(\partial E/\partial k_z)_{\mathscr{A}_k}. \tag{1.31}$$

Clearly in fig. 13(d) it is the difference between the shaded areas that gives Z and $\bar{v}_z$.

8. At an extremum of $\mathscr{A}_k$ with respect to k_z (a k-orbit which controls the periodicity of the de Haas–van Alphen and related effects),[19] $Z = 0$ and the orbit in real space is closed, though not in general plane. According to Ampère's theorem an electron in such an orbit produces the same magnetic field pattern as a magnetic shell having edges that coincide with the r-orbit, and a magnetic moment per unit area of $\omega_c e/2\pi$, being the current flowing round the orbit. Since, according to (26) the orbit has area $\mathscr{A}_k/\alpha^2$ when projected onto the transverse plane, the component of its total moment, parallel to **B**, is

$$\mu_z = \omega_c e\mathscr{A}_k/2\pi\alpha^2 = [B(\partial \ln \mathscr{A}_k/\partial E)_{k_z}]^{-1} \tag{1.32}$$

The transverse components of μ may also be related to the geometry of the energy surfaces,[20] but we shall not have occasion to use them.

Historical note

The expression (29) relating ω_c to the geometry of the energy surfaces, without an explicit algebraic representation of their shape, was a pioneer example of a development of great importance to the study of real metals, marking an advance on the analytical models which had come into vogue. Shockley's[21] discussion of low-field magnetoresistance in terms of what he called 'tube-integrals' included something like (29), and at the same time Onsager and Lifshitz[18] were developing their geometrical theories of the de Haas–van Alphen effect which gave immense impetus to the new science of Fermiology – mapping out the Fermi surface of individual metals as the first stage to characterising them and explaining the variety of behaviour shown by different metals.

The term Fermi surface has already been introduced as the energy surface in k-space for $E = E_F$. The first use of the term is obscure. It appears in 1952 in papers by Onsager[22] and Chambers,[23] its meaning taken for granted and used so nonchalantly as to suggest it had been accepted in conversational usage. Shoenberg uses it *en passant* in his unpublished Ph.D. dissertation of 1935, and it may therefore have entered the local Cambridge patois already.

Effective path and DC conductivity

The formula (1) is valid only for an isotropic electron gas when $\mathbf{B} = 0$. We now derive a general expression to cover any shape of Fermi surface and any scattering process, in an arbitrary uniform magnetic field.

It is convenient to treat the steady applied electric field $\mathscr{E}$ as a rapid sequence of pulses which have been applied at all times until the present. Since we are concerned only with the linear response to weak fields, the response to the steady field is the sum of all impulse responses, each of which is treated as consisting of $\mathscr{E}$ applied for a time δt to the electron gas in equilibrium. During this short interval each electron, in accordance with (24), suffers a change $\delta k = e\mathscr{E}\delta t/\hbar$, so that the distribution of filled states is shifted bodily in k-space, as in fig. 14. As already remarked, it is only states close to the Fermi surface that have their occupation affected, and we may imagine the effect of the impulse to be the creation of new electrons at some parts of the surface and annihilation of an equal number elsewhere. To calculate the number we take the division between occupied and unoccupied states to be sharp, as at 0 K; raising the temperature does not alter the number, but only spreads them over a few $k_B T$ in energy. Any elementary area δS on the Fermi surface sweeps out a volume $\delta S \cdot \delta k$, containing $\delta S \cdot \delta k/4\pi^3$ electrons per unit volume of metal. We therefore write for the number created between t' and $t' + \delta t'$

$$\delta^2 n = e\mathscr{E} \cdot \delta S \delta t'/4\pi^3 \hbar. \tag{1.33}$$

Since each new electron starts with the Fermi velocity $\mathbf{v}_F$ appropriate to the

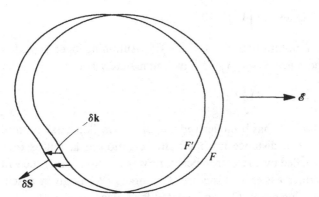

Figure 1.14 Electrons with $\mathbf{k}$ lying within the Fermi surface F are displaced by an impulsive field to points δk away, and lie within F'. An element of F' is represented by δS, directed normal to the surface and pointing towards state of higher energy.

element δS, the initial current density due to these elements is

$$\delta^2 \mathbf{J}(t') = e^2 \mathbf{v}_F (\mathscr{E} \cdot \delta \mathbf{S}) \delta t' / 4\pi^3 \hbar. \tag{1.34}$$

Let us now watch this current evolve under the influence of $\mathbf{B}$ and of scattering processes. It may change direction so that any one component oscillates, but eventually it will die out. A bunch of electrons, all generated opposite the same δS, will begin moving as a compact bunch, but scattering will disperse them into different directions so that the velocity of their centroid, $\mathbf{v}$, initially $\mathbf{v}_F$, falls to zero. No matter when t' may have been, $\mathbf{v}$ at $t' + \Delta t$ is determined by Δt alone: $\mathbf{v} = \mathbf{v}(\Delta t)$. Thus the electrons created at $t'(< t)$ have a mean velocity $\mathbf{v}(t - t')$ at time t, and if $\mathbf{v}$ is substituted for $\mathbf{v}_F$ in (34) the resulting $\delta^2 \mathbf{J}$ is the contribution to $\mathbf{J}$ from the field applied between t' and $t' + \delta t'$. To find the effect of a steady field we need only to integrate t' from $-\infty$ to t, so that

$$\delta \mathbf{J} = (e^2 \mathscr{E} \cdot \delta \mathbf{S} / 4\pi^3 \hbar) \int_0^\infty \mathbf{v}(\Delta t) \, \mathrm{d}(\Delta t)$$

$$= (e^2 \mathscr{E} \cdot \delta \mathbf{S} / 4\pi^3 \hbar) \mathbf{L}, \tag{1.35}$$

in which $\mathbf{L}$ represents the integral and is the mean vector distance travelled by each electron in the bunch from the moment of its creation until scattering brings the centroid to rest. $\mathbf{L}$ is the effective path and is, of course, a function of position on the Fermi surface, of $\mathbf{B}$ and of the scattering mechanisms at work.

Finally, we integrate over the Fermi surface to obtain the total current due to a steady field $\mathscr{E}$,

$$\mathbf{J} = (e^2 / 4\pi^3 \hbar) \int \mathbf{L}(\mathscr{E} \cdot \mathrm{d}\mathbf{S}). \tag{1.36}$$

In subscript notation $\mathbf{L}(\mathscr{E} \cdot \mathrm{d}\mathbf{S}) = L_i \mathscr{E}_j \mathrm{d}S_j$ (summing over repeated subscripts), and since $J_i = \sigma_{ij} \mathscr{E}_j$ we have immediately that

$$\sigma_{ij} = (e^2 / 4\pi^3 \hbar) \int L_i \mathrm{d}S_j. \tag{1.37}$$

This basic formula has been derived in many ways. Some like to think of $\mathbf{L}$ not as the mean distance travelled after creation at $\mathbf{k}$, but as the mean distance travelled by an electron before its wave-number takes the value $\mathbf{k}$. I find the former easier to visualize. Chambers[24] has given a scholarly review of the history of (37) and related formulae.

Fermi surfaces are centro-symmetric, i.e. $E_F(-\mathbf{k}) = E_F(\mathbf{k})$, though not infrequently $\pm \mathbf{k}$ lie on different surfaces which form a centro-symmetric pair. Corresponding elements at $\pm \mathbf{k}$ have both $\mathrm{d}\mathbf{S}$ and $\mathbf{L}$ with opposite signs, contributing equally to σ_{ij}. And there may be other elements related to these by symmetry operations; in a cubic crystal, for example, an arbitrary **k** is in general one of a set of 48 equivalents, but only so long as **B** = 0. When **B** ≠ 0 different elements of the set make different angles with **B** and the equivalence is lost.

In zero magnetic field the electrons travel in straight lines between collisions. When impurities or other defects smaller than the electron wavelength are the dominant source of collisions, the electron may be deviated through so large an angle at each process that it loses all sense of its previous motion. Such scattering is called catastrophic and may be considered to bring the electron to a halt. In this case the original bunch continues without it in the same direction, the number in the bunch falling exponentially; **L** is now parallel to v_F and has the same magnitude as the usual mean free path, *l*. There may, however, be larger defects which mainly scatter electrons through small angles, many processes being required to randomize the motion of each. The mean free path may now prove to be a less generally useful concept, in the sense that a different value is required for each phenomenon being studied, but **L** is still well defined. There is no intrinsic reason why **L** should be exactly parallel to v_F, for if v_F does not lie along a symmetry direction the multitude of small-angle processes may bias **v** towards or away from the symmetry direction, and **L** will be correspondingly biased. In a cubic crystal symmetry demands that the component of **L** transverse to v_F shall be compensated when all elements of the set are taken together, and we may therefore interpret **L** as the component of the effective path that is parallel to v_F. In non-cubic crystals this argument fails, but it is not important for anything considered in this book.

Completing the story as it concerns cubic crystals when **B** = 0, we note that σ_{ij} is isotropic and diagonal. Choosing any direction at will, we write (37) in the form

$$\sigma = (e^2/4\pi^3\hbar) \int L dS \cos^2 \xi, \tag{1.38}$$

where ξ is the angle between **L** (and d**S**) and the chosen direction. In particular we may take the cube axes of the crystal as coordinate axes, and the *x*-axis as defining the chosen direction, when $\cos \xi$ becomes one of the three direction-cosines of **L**, (c_1, c_2, c_3) where $c_1^2 + c_2^2 + c_3^2 = 1$. The 48 equivalent directions are obtained as permutations of c_1, c_2 and c_3 with either sign; when all are taken together they cause $\cos^2 \xi$ to have an average value of $\frac{1}{3}$, so that

$$\sigma_0 = e^2 \bar{L} S/12\pi^3\hbar, \tag{1.39}$$

$\bar{L}$ being the average of L over the whole Fermi surface of area S.[23]

Comments on the conductivity formula

1. *Application to a free-electron gas with catastrophic scattering*

As illustration we calculate **L** for a free-electron gas and its variation with **B**, assuming catastrophic scattering. Consider a plane section δk_z thick, cut normal to **B**, and having radius k_t, where $k_t^2 + k_z^2 = k_F^2$. In r-space each electron describes a helical path of radius $r_t = k_t/\alpha$ and moves along **B** with speed $k_z v_F/k_F$. We shall, however, ignore the z-component which is unaffected by **B**. The transverse motion is represented in fig. 15 which is to be imagined a complex plane so that the position of the electron is described by $(k_t/\alpha)\,e^{i\phi}$.

Of a bunch of electrons starting at P a fraction, $\exp(-\phi/\omega_c\tau)$, reach ϕ without colliding, and $(d\phi/\omega_c\tau)\exp(-\phi/\omega_c\tau)$ collide in the next $d\phi$ and are on the average brought to rest there. The position of the centroid when all have collided is therefore at ϕ_0, where

$$e^{i\phi_0} = \int_0^\infty e^{i\phi}(d\phi/\omega_c\tau)\,e^{-\phi/\omega_c\tau} = 1/(1 - i\omega_c\tau). \qquad (1.40)$$

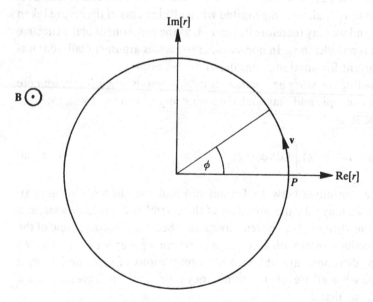

Figure 1.15 Notation to describe free-electron motion on a complex-plane in r-space.

For these electrons, then.

$$L = (k_t/\alpha)(e^{i\phi_0} - 1) = (k_t/\alpha)i\omega_c\tau/(1 - i\omega_c\tau). \qquad (1.41)$$

Obviously electrons which start at ϕ' execute the same variety of paths, but turned through ϕ', so that for them this expression must be multiplied by $e^{i\phi'}$.

Now if $\mathscr{E}$ lies along the real axis, the electrons starting at P have their initial velocity normal to $\mathscr{E}$, and those starting at ϕ' make an angle $\pi/2 + \phi'$. For $\mathscr{E} \cdot d\mathbf{S}$ in (36) we write $- k_t\mathscr{E} \sin \phi' \, d\phi' \delta k_z$, and evaluate J to give

$$\delta\sigma = - (e^2/4\pi^3\hbar)(k_t^2/\alpha)[i\omega_c\tau\delta k_z/(1 - i\omega_c\tau)] \int_0^{2\pi} \sin \phi' \, e^{i\phi'} \, d\phi'$$

$$= e\omega_c\tau \cdot \delta n/B(1 - i\omega_c\tau)$$

where $\delta n = \pi k_t^2 \delta k_z/4\pi^3$, the number density of electrons in the slice. Hence for the whole electron gas, by use of (3), we have

$$\sigma = \sigma_0/(1 - i\omega_c\tau)$$

in agreement with (20).

2. Kohler's rule

A bunch of electrons, all having the same initial $\mathbf{k}$, executes an orbit whose linear dimensions are inversely proportional to B, and is dissipated by collisions. If B is increased by a factor a, while the collision rate is similarly increased, the pattern of electron behaviour is simply scaled down without change of character; but only if the extra collisions are of the same sort – more impurities, or more phonons with the same frequency distribution (if that is possible). Under these conditions L and all components of σ_{ij} are reduced by a, and any measured resistivity increased by a. It follows that keeping the ratio B/ρ_0 constant ensures that in different samples the probability of being scattered in the course of one cycle of the orbit remains constant, and the relative effect of B on the resistivity should be the same for all –

$$\Delta\rho/\rho_0 = F_1(B/\rho_0) \quad \text{or} \quad F_2(r_0 B) \qquad (1.42)$$

or, where appropriate,

$$\Delta\rho/\rho_0 = F_3(\omega_c\tau). \qquad (1.43)$$

In these expressions $\Delta\rho$ stands for $\rho(\mathbf{B}) - \rho_0$, and may be taken to apply to any component of ρ_{ij}, though the dependence on B/ρ_0 will be different for each. This is Kohler's[25] rule and an example that illustrates its validity is to

9   be found in fig. 4. At higher values of $\omega_c\tau$ different samples give scattered points.[26]

The samples of indium used for fig. 4 differed primarily in their impurity content, so that the condition under which the rule was derived was probably quite well satisfied. This is not so when phonon scattering dominates, and ρ_0 is changed by raising the temperature, for the wavelength of the most important phonons falls in the process, and with it the scattering

126   pattern is altered. Fig. 3.28 shows how the results at different temperatures do not coincide with a unique curve on the Kohler plot of $\Delta\rho/\rho_0$ against $\omega_c\tau$. Kohler's rule also breaks down when the effects of orbit quantization

12   are observable (fig. 10), since the quantum number (the number of wavelengths in a complete orbit) is changed with B and cannot be kept fixed at the same time as B/ρ_0.

When ρ saturates, as in curve (*a*) of fig. 5, $F_1(B/\rho_0)$ in (42) becomes independent of B and therefore of ρ_0; the saturation ratio ρ_∞/ρ_0 is the same for all samples. On the other hand, in curve (*b*), $\Delta\rho/\rho_0 \propto B^2$ and (42) takes the form

$$\Delta\rho/\rho_0 = \text{const.}\,(B/\rho_0)^2. \tag{1.44}$$

In a given field $\Delta\rho$ is inversely proportional to ρ_0 and the purest samples rise to the highest resistivities. Finally we may note that if, as in fig. 6, $\rho \propto B$, the high-field resistivity becomes virtually independent of ρ_0 and the state of purity of the sample.

3. *The magnetoresistance is normally positive*[27]

The following argument is closely related to a well-known circuit theorem to the effect that if the total current flowing through a network of linear resistors is fixed, the current distributes itself among the branches so as to minimize the heat production. Similarly when a metal carries a given current density, the displacement of the different parts of the Fermi surface when $\mathbf{B} = 0$ is such as to minimize the dissipation rate by scattering.

Let us first analyse the energy balance when a steady $\mathbf{J}$ is accompanied by a steady field $\mathscr{E}_0$, not necessarily parallel to $\mathbf{J}$. An element δS, whose normal is set at an angle ψ to $\mathscr{E}_0$, is displaced along its normal by k_n, say, where (when $\mathbf{B} = 0$),

$$k_n = p\mathscr{E}_0 \cos\psi, \tag{1.45}$$

p being a function of position on the Fermi surface. This displacement is maintained by the balance of the creation rate for new particles, obtained from (33):

$$\delta \dot{n} = e \mathscr{E}_0 \delta S \cos \psi / 4\pi^3 \hbar, \tag{1.46}$$

and an equal rate of scattering of particles into other regions of k-space near the Fermi surface. It does not matter whether their motion is randomized by elastic or inelastic collisions; in the former case their excess energy is distributed within the electron gas, making it hotter, but eventually it will find its way by inelastic processes to the lattice.

The rate at which energy is fed by $\mathscr{E}_0$ into the electrons in the vicinity of δS is

$$\delta \dot{W}_0 = \mathscr{E}_0 \delta J \cos \psi = e \mathscr{E}_0 v k_n \delta S \cos \psi / 4\pi^3, \quad \text{from (34),} \tag{1.47}$$

and this is also the rate at which energy is carried away by the scattered electrons. Comparison of (46) and (47) shows that each electron that is scattered out takes with it $\hbar k_n v$, which is just the excess energy of an electron displaced k_n from the Fermi surface. By this argument we make plausible the assumptions involved in the next stage.

When **B** is applied the movement round the Fermi surface under the influence of the Lorentz force alters the distribution of extra particles, so that k_n in (45) and (47) must be replaced by $k_n + k'_n$, and of course $\mathscr{E}$ is changed. Let us first eliminate $\mathscr{E}_0$ from (47) by use of (45) to express the dissipation rate in terms of k_n, however that k_n may have arisen:

$$\delta \dot{W}_0 = e v k_n^2 \delta S / 4\pi^3 p.$$

When k_n is supplemented by k'_n the dissipation is changed to

$$\delta \dot{W} = e v (k_n + k'_n)^2 \delta S / 4\pi^3 p$$

and for the whole Fermi surface,

$$\dot{W} = (e/4\pi^3) \int v \, dS (k_n + k'_n)^2 / p$$

$$= \dot{W}_0 + (e/4\pi^3) \int (v \, dS/p)(2 k_n k'_n + k'^2_n)$$

$$= \dot{W}_0 + (e \mathscr{E}_0 / 2\pi^3) \int v k'_n \, dS \cos \psi + (e/4\pi^3) \int v k'^2_n \, dS / p.$$

The first term is the dissipation when $B = 0$ and the second, obtained by use of (45), differs only in its coefficient from the extra contribution by k'_n to the current parallel to the original direction of $\mathscr{E}_0$; since **J** is kept constant this term vanishes. The third term is essentially positive, so that $\dot{W} \geqslant \dot{W}_0$ and the resistivity is increased, or unchanged if k'_n is everywhere zero as in a free-electron gas.

As with Kohler's rule there are exceptions, especially in ferromagnetic metals where the scattering can be affected by the spin alignment produced by B (fig. 7). Where boundary scattering is important B may deflect electrons towards or away from the boundary and thus alter its influence (fig. 8); and, as usual, one must not rely on essentially classical arguments when orbit quantization is significant, though examples of negative magnetoresistance arising in this way are rare in the literature, if indeed there are any.

4. Onsager's reciprocal relations

Onsager[28] showed that when a system initially in equilibrium is disturbed by applied forces, or analogues of force, F_i, so that there arise fluxes of energy, matter, etc. Φ_i that are linearly related to the forces ($\Phi_i = \mathscr{L}_{ij}F_j$), then the coefficients $\mathscr{L}_{ij}$ obey certain symmetry relations. It should be noted that the F_i and Φ_i must be defined in strict accordance with prescription, but if this is done $\mathscr{L}_{ij}(B) = \mathscr{L}_{ji}(-B)$. For a system that is unaffected by magnetic field the tensor $\mathscr{L}_{ij}$ is symmetrical. There is no need to expound the general theory here – it is enough to know that $\sigma_{ij}(B) = \sigma_{ji}(-B)$, an example of which is found in (18).

It is easy to demonstrate this result when scattering is catastrophic. On a thin slice of the Fermi surface cut normal to B describe two elements of area δS_l and δS_m, proportional to the electron velocities there,

$$\delta S_l / v_l = \delta S_m / v_m, \tag{1.48}$$

and let the normals to the surface at these two points be defined by unit vectors $\mathbf{n}_l$ and $\mathbf{n}_m$, parallel to the velocity vectors. Now apply an impulsive field of unit strength, defined by the vector $\mathbf{n}_0$, which according to (33) will create $(e\delta S_l / 4\pi^3 \hbar)\mathbf{n}_0 \cdot \mathbf{n}_l$ extra electrons next to δS_l. They will migrate round the Fermi surface, and if they suffer no collisions will, according to Liouville's theorem, in due course fully occupy the region next to δS_m; this is ensured by (48). Here they will generate a current with a component in the direction $\mathbf{n}_1$,

$$\delta J = p_{lm}(e^2 v_m \delta S_l / 4\pi^3 \hbar)(\mathbf{n}_0 \cdot \mathbf{n}_l)(\mathbf{n}_1 \cdot \mathbf{n}_m). \tag{1.49}$$

In this expression, p_{lm} has been inserted to represent the fraction of electrons which leave δS_l and reach δS_m without suffering collisions.

By repeating this process with all other elements on the Fermi surface, associating with each δS_l a corresponding δS_m reached from it in the same time, and then integrating (49), we find an expression for the impulse response of $\mathbf{J} \cdot \mathbf{n}_1$ at a certain time after an impulse $\mathscr{E}$ parallel to $\mathbf{n}_0$.

Now repeat the process with $\mathbf{B}$ reversed and with $\mathscr{E}$ applied along $\mathbf{n}_1$. The electrons now move in the opposite sense, so that those generated opposite δS_m reach δS_l in the same time as before, and with the same chance p_{lm} of arriving without collision on the way. When the current along the direction $\mathbf{n}_0$ is written down it differs from (48) only in that $v_m \delta S_l$ is replaced by $v_l \delta S_m$ which takes the same value according to (48). The same reciprocity applies to every element and every time interval; consequently we may say that if $\mathscr{E}\mathbf{n}_0$ sets up a current density $J\mathbf{n}_1$, $\mathscr{E}\mathbf{n}_1$ will set up a current density $J\mathbf{n}_0$ but only if $\mathbf{B}$ is reversed in sign. This is the demonstration of Onsager's reciprocity relation in the particular case of interest; it is, of course, much more general and, for instance, applies to all types of scattering. Even if the scattering process is $\mathbf{B}$-dependent it holds, though one proviso is necessary – the material must attain thermodynamic equilibrium before the electric field is applied. This is implicit in Onsager's derivation, but it is easy to forget that a ferromagnet showing hysteresis in its magnetization curve is not in equilibrium, and reversing $\mathbf{B}$ may set up a quite different microscopic structure; reciprocity then fails.

5. *Hall effect in a strong field*

The argument that follows is beguilingly simple, and it should not be forgotten that it is valid only for the limiting behaviour of σ_{xy} and other off-diagonal components of the conductivity. Diagonal components demand a deeper analysis which will be deferred to chapter 3. Let us consider the motion of electrons on a slice in k-space, remembering that (26) allows us to translate the results into r-space. An electron created at P (see fig. 11) may make many circuits of its orbit before being scattered, and one may suppose that the final position of the centroid of a bunch starting at P will hardly depend on where P lies on the orbit. The stronger $\mathbf{B}$ is, the less important does the initial point become.

Without detailed knowledge of the dynamics and scattering behaviour of the electrons we cannot find the final position Z – unless the scattering is catastrophic it need not even lie in the same plane. For the Hall effect none of this matters and we place Z anywhere plausible and suppose it marks the projection on the plane of the real final position; we take Z as the origin of coordinates for the slice. Translated into real space the diagram represents the effective path $\mathbf{L}$ (strictly, its component $\mathbf{L}_t$ normal to $\mathbf{B}$) as a vector of length α^{-1} times PZ turned through $\pi/2$. Alternatively $\mathbf{k}_t = -\boldsymbol{\alpha} \wedge \mathbf{L}_t$.

From (37), disregarding signs,

$$\sigma_{xy} = (e^2/4\pi^3\hbar) \int L_x dS_y = (e/4\pi^3 B) \int k_y dS_y.$$

Since dS is directed normal to the Fermi surface, dS_y for a slice δk_z thick can be written as $\delta k_z dk_x$, and the integral becomes a line integral, $\oint k_y dk_x$, which is just $\mathscr{A}_k$, the area of the k-orbit. Then, since $\mathscr{A}_k \delta k_z/4\pi^3$ is the number of electrons per unit volume whose k lies in the slice, we may integrate over the whole Fermi surface:

$$\sigma_{xy} = ne/B. \tag{1.50}$$

We shall see later that frequently σ_{xx} and σ_{yy} vary in high fields as $1/B^2$, and then, as (14) shows and the free-electron gas illustrates,

$$\rho_{yx} \sim 1/\sigma_{xy} \sim B/ne, \tag{1.51}$$

so that the limiting form of the Hall constant is

$$R_H \sim 1/ne, \tag{1.52}$$

the same as for a free-electron gas, even though the energy surfaces may be entirely different. All we ask is that they be closed, so that $\mathscr{A}_k$ has a meaning for all sections.

The expression (52) applies at all values of **B** to a free-electron gas, but in general only in the limit of high **B**; in the next section we shall meet an example where it describes the high-field behaviour well but gives even the sign of the Hall field incorrectly when **B** is smaller. There are many ways of deriving (52),[29] some of which make use of the fact that the result is independent of scattering, which may therefore be ignored, if convenient. In fields $\mathscr{E}_x$ and B_z any charged particle has drift velocity $v_y = \mathscr{E}_x/B_z$ superimposed on its orbital motion, and (50) is an expression of this behaviour. But there is no need to go further into alternative approaches to the simplest of the problems in magnetoresistivity.

Elementary extensions of the free-electron theory

1. Electrons and holes

The concept of a hole, with properties that simulate a positively charged electron, is too well established in semiconductor physics to need careful exposition at this point. All the same, its use in connection with metals is sufficiently different in some respects from what is usually understood that we must discuss it in more detail at the right time, in chapter 3. It is enough at present to consider a metal in which there are two independent types of particle, both behaving classically, and differing only in charge, i.e. n_- electrons and n_+ holes per unit volume. If each were present alone, it would confer positive conductivity on the metal in zero magnetic field, σ_- for the electrons and σ_+ for the holes; and in the presence of a transverse field **B** each would behave according to (16), or (20) if complex notation is used. Thus for the two together,

$$\sigma = \sigma_-/(1 - i\gamma_-) + \sigma_+/(1 + i\gamma_+), \tag{1.53}$$

in which γ is written for $|\omega_c\tau|$ and the opposite charges are reflected in the signs in the denominators. By use of (3) we rewrite this:

$$\sigma = (n_-e/B)[\gamma_-/(1 - i\gamma_-) + c\gamma_+/(1 + i\gamma_+)] \tag{1.54}$$

in which $c = n_+/n_-$. If the density of holes equals that of electrons, $c = 1$ and the metal is said to be compensated.

The resistivity ρ is $1/\sigma$ and looks more complicated:

$$\rho = \frac{B}{n_-e} \frac{[\gamma_- + c\gamma_+ + \gamma_-\gamma_+(\gamma_+ + c\gamma_-)] - i[\gamma_-^2 - c\gamma_+^2 + (1 - c)\gamma_-^2\gamma_+^2]}{(\gamma_- + c\gamma_+)^2 + (1 - c)^2\gamma_-^2\gamma_+^2}. \tag{1.55}$$

The low and high field limits are easily found by keeping only terms up to $\gamma^2(\propto B^2)$ for the former, and only the highest orders in γ for the latter. Then, writing ρ_0 doe $(\sigma_- + \sigma_+)^{-1}$, and ρ_1 for ρ_{xx}, the real part of ρ, we have

$$(\rho_1/\rho_0)_{\text{low}} \sim 1 + c\gamma_-\gamma_+(\gamma_- + \gamma_+)^2/(\gamma_- + c\gamma_+)^2, \tag{1.56}$$

describing a quadratic magnetoresistance, $\Delta\rho_1/\rho_0 \propto B^2$. Unless $c = 1$ the increase is not continued indefinitely, but eventually saturation occurs at a value ρ_∞:

$$\rho_\infty/\rho_0 \sim A/(1 - c)^2, \quad \text{where } A = (\sigma_- + \sigma_+)(1/\sigma_- + c^2/\sigma_+). \tag{1.57}$$

In a compensated metal (55) simplifies, by use of (54) and without approximation, to a pure quadratic effect:

$$(\rho_1/\rho_0)_{\text{comp}} = 1 + \gamma_-\gamma_+. \tag{1.58}$$

When compensation is not quite perfect, so that $c = 1 - \varepsilon$ ($\varepsilon \ll 1$), the high field limit (57) is of the order of $4/\varepsilon^2$, which may be very large but is not reached until B itself is large. The quadratic rise described by (58) would need to continue until $\gamma \sim 2/\varepsilon$ before it met the saturation value. This can be illustrated by choosing $\gamma_+ = \gamma_- = \gamma$ in (55) when, without approximation,

$$\rho_1/\rho_0 = (1 + \gamma^2)/[1 + \gamma^2\varepsilon^2/(1 + c)^2]. \tag{1.59}$$

Examples are shown in fig. 16. When ε is small the initial stages of the curve

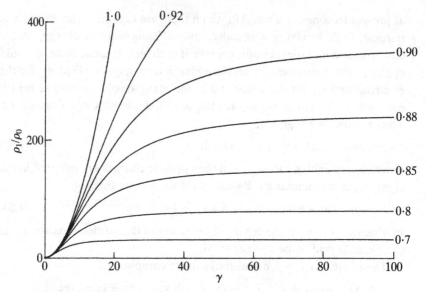

Figure 1.16 Transverse magnetoresistance of a metal containing both electrons and holes, according to (59); values of c, i.e. n_+/n_-, are given beside each curve.

9   differ little, following (58), so that $\Delta\rho_1/\rho_0 \sim (\omega_c\tau)^2$ as has already been noted as an empirical relation. When, as in bismuth,[6] the quadratic rise has been found to continue until $\rho_1/\rho_0 > 10^6$, ε must be less than 10^{-3}; with only about 10^{-5} electrons per atom, the material must have been pure enough for the numbers of electrons and holes to differ by, at most, 10^{-8} per atom. As is well known in semiconductor physics, this degree of purity demands great attention to sample purification.

The same is true for the Hall effect, as described by the imaginary part of (55). Unless $c = 1$ the high-field limit is

$$\rho_{yx} \sim B/ne(1 - c) = B/e(n_- - n_+) \qquad (1.60)$$

which is a generalization of (51). When $\varepsilon \ll 1$ a high field is needed for this limit to be reached; until this is achieved (and always when $c = 1$), terms of lower order in γ must be kept when approximating to (55). If $c = 1$, without approximation,

$$\rho_{yx} = (B/n_-e)(\gamma_- - \gamma_+)/(\gamma_- + \gamma_+). \qquad (1.61)$$

In bismuth the electrons are considerably more mobile than the holes, and $\gamma_- \gg \gamma_+$. In a perfectly compensated sample, then, one expects from (61) that the sign of the Hall effect will reflect the dominance of electrons in the conduction process. A not-very-pure sample, however, in which holes slightly outnumber electrons must ultimately settle down in the form determined by (60) with the Hall field of opposite sign. In the end numbers are more important than mobility. Reversal of the sign of the Hall effect is not uncommon in bismuth samples.[30] It is always found in aluminium,[31] which is not compensated and has more holes than electrons; it is this fact that controls the high-field behaviour. At low fields, however, it behaves more like a free-electron metal with three electrons per atom. The simple model used here is now inadequate, and the same must be said of the curiosities of Hall 'constant' in magnesium, zinc and other metals suffering magnetic breakdown. In later chapters these matters will receive more realistic appraisal.

In an uncompensated metal the Hall angle, φ in fig. 1, rises to $\pi/2$, while in a compensated metal it tends to remain small and eventually falls to zero. This behaviour follows immediately from the high-field expressions (57) and (60) for the uncompensated metal, and (58) and (61) for the compensated. In the former $\tan \varphi \propto B$, in the latter $\tan \varphi \propto 1/B$. The Hall effect may be hard to measure in a compensated metal since the transverse component of $\mathscr{E}$ is commonly much smaller than the longitudinal. The reverse holds in an uncompensated metal, where the Hall field may be large and create difficulties in measuring the longitudinal component and hence the magnetoresistivity. Techniques for overcoming the problems, and the new problems raised by the techniques themselves, are the subject of chapter 2.

2. *Anisotropic scattering*

The simple argument incorporated in fig. 1 depends for its validity on the relaxation time being independent of the direction of **J**. Since currents in different directions involve different displacements of the Fermi surface, variations of scattering rate over the surface are likely to cause anisotropy in the relaxation rate of **J**, unless crystal symmetry so constrains those variations that σ_{ij} must be isotropic. Most cases of anisotropic scattering require long and tedious analysis to work out fully, but a single simplified model will serve to show that the effects to be expected are rather insignificant in comparison with most of the examples of interest, e.g. figs. 5, 6 and 10. Even with the most drastic simplification it is still desirable to concentrate on one feature only, the ratio ρ_∞/ρ_0.

Consider then a two-dimensional free-electron gas for which the probability of an electron being scattered catastrophically in time δt varies with its direction of motion, ϕ, as $(p_0 + p_1 \cos 2\phi)\delta t$, p_0 and p_1 being constants ($p_1 < p_0$). The principal axes of the zero-field conductivity are at $\phi = 0$, the x-axis, and $\phi = \pi/2$, the y-axis; $\sigma_{yy} \neq \sigma_{xx}$. Let us first calculate these zero-field conductivities, beginning with $\sigma_{xx}(0)$. An electron starting out at an angle ϕ moves in a straight line and its effective path is $v_F \tau(\phi)$, where $\tau = (p_0 + p_1 \cos 2\phi)^{-1}$. Then $L_x = v_F \cos \phi/(p_0 + p_1 \cos 2\phi)$ and from (37), suitably modified for two dimensions,

$$\sigma_{xx}(0) = C \int_0^{2\pi} \cos^2 \phi \, d\phi/(p_0 + p_1 \cos 2\phi)$$
$$= (\pi C/p_1)[1 - (p_0 - p_1)^{1/2}/(p_0 + p_1)^{1/2}].$$

Here C is a constant whose value may be determined by inspection. Since $\sigma_{xx}(0)$ must equal ne^2/mp_0 when $p_1 = 0$, $C = ne^2/\pi m$ and

$$\sigma_{xx}(0) = (ne^2/mp_1)[1 - (p_0 - p_1)^{1/2}/(p_0 + p_1)^{1/2}]. \tag{1.62}$$

Similarly,

$$\sigma_{yy}(0) = (ne^2/mp_1)[(p_0 + p_1)^{1/2}/(p_0 - p_1)^{1/2} - 1]. \tag{1.63}$$

Now we turn to the high-field behaviour, for which we need to know $L(\phi_0)$ for an electron starting at an angle ϕ_0. Out of a bunch starting together the fraction f reaching ϕ without being scattered is given by the solution of

$$df/d\phi = -f(p_0 + p_1 \cos 2\phi)/\omega_c, \tag{1.64}$$

i.e.

$$f = \exp\{-p_0(\phi - \phi_0)/\omega_c - \tfrac{1}{2}p_1(\sin 2\phi - \sin 2\phi_0)/\omega_c\}. \tag{1.65}$$

For the high-field behaviour, we note that when ω_c/p_0 is large only a small fraction, $2\pi p_0/\omega_c$, of the electrons are scattered during their first orbit and we may find the centroid of the points where scattering events occur by expanding (64) and stopping after the term in $1/\omega_c$. Those electrons that continue and are scattered in subsequent orbits have the same centroid, which is therefore the final position of all electrons starting from ϕ_0. To concentrate on L_x alone, we have

$$L_x \sim - \int_{\phi_0}^{\phi_0 + 2\pi} (k_F/\alpha)(\sin \phi - \sin \phi_0)(df/d\phi) \, d\phi/(2\pi p_0/\omega_c), \tag{1.66}$$

in which $(k_F/\alpha)(\sin\phi - \sin\phi_0)$ is the distance along x travelled by an electron which started from ϕ_0 in an orbit of radius k_F/α and collided at ϕ. In this limit we replace (64) by

$$df/d\phi \sim -[1 - p_0(\phi - \phi_0)/\omega_c$$
$$- \tfrac{1}{2}p_1(\sin 2\phi - \sin 2\phi_0)/\omega_c](p_0 + p_1 \cos 2\phi)/\omega_c. \tag{1.67}$$

Now when we seek to find σ_{xx} we must evaluate $\int L_x \cos \phi_0 \, d\phi_0$, to which only those terms in L_x that vary as $\cos \phi_0$ can contribute. We may disregard all else in (66) and write for the expurgated effective path

$$`L_x' = (k_F/\alpha\omega_c)(p_0 - \tfrac{1}{2}p_1)\cos\phi_0, \quad \text{being proportional to } 1/B^2.$$
(1.68)

Hence at high fields, by the same process as led to (62),

$$\sigma_{xx} \sim C'(p_0 - \tfrac{1}{2}p_1)/B^2$$

and we must set $C' = nm$ so that the result will agreed with (16) when $p_1 = 0$. Therefore,

$$\sigma_{xx} \sim nm(p_0 - \tfrac{1}{2}p_1)/B^2 \quad \text{and} \quad \sigma_{yy} \sim nm(p_0 + \tfrac{1}{2}p_1)/B^2. \quad (1.69)$$

From (50), $\sigma_{xy} = ne/B$, and it follows from (14) that

$$\rho_{xx}(\infty) = (m/ne^2)(p_0 + \tfrac{1}{2}p_1) \quad \text{and} \quad \rho_{yy}(\infty) = (m/ne^2)(p - \tfrac{1}{2}p_1),$$

or, from (62) and (63)

$$\left.\begin{array}{l} \rho_{xx}(\infty)/\rho_{xx}(0) = \rho_{xx}(\infty)\sigma_{xx}(0) \\ \qquad = (r + \tfrac{1}{2})[1 - (r-1)^{1/2}/(r+1)^{1/2}] \end{array}\right\} \quad (1.70)$$

and

$$\rho_{yy}(\infty)/\rho_{yy}(0) = (r - \tfrac{1}{2})[(r+1)^{1/2}/(r-1)^{1/2} - 1]$$

in which $r = p_0/p_1$. These results are plotted in fig. 17; the magnetoresistance is positive, as expected, and is more pronounced when $\mathbf{J}$ lies along y,

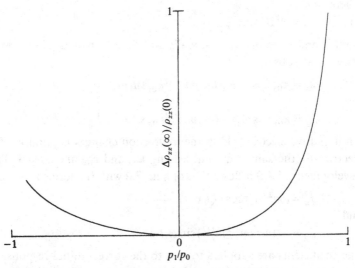

Figure 1.17 Transverse magnetoresistance resulting from anisotropic scattering, according to (70).

the better-conducting direction, for **B** then brings electrons into the regions where scattering is stronger. However, the effect is weak – a factor of 10 in scattering rate between $\phi = 0$ and $\phi = \pi/2$ is equivalent to $r = 1.22$ and leads to saturation of ρ_{yy} 56% above the value when $B = 0$. It only needs $c = 0.11$ in (59), i.e. 11% of holes added to an electron gas, to produce the same result.

3.   *Anisotropy of effective mass*

If the electrons are quasi-free, but with different effective masses along the three principal axes – a situation that arises in bismuth and in some semiconductors – the energy surfaces are ellipsoids,

$$\tfrac{1}{2}\hbar^2 (k_x^2/m_x + k_y^2/m_y + k_z^2/m_z) = E. \tag{1.71}$$

We now show that, provided τ is the same for all directions of motion, this anisotropy produces no magnetoresistance. One can show this by carrying through the same argument as relates to fig. 1, but it may be more convincing if demonstrated in detail, from the impulse response.

From (24), the equations of motion can be written

$$\dot{k}_x = \alpha v_y = eBk_y/m_y \quad \text{and} \quad \dot{k}_y = -eBk_x/m_x, \tag{1.72}$$

so that

$$\ddot{k}_x + \omega_c^2 k_x = 0 \quad \text{and} \quad \ddot{k}_y + \omega_c^2 k_y = 0,$$

where

$$\omega_c^2 = e^2 B^2/m_x m_y.$$

Both k_x and k_y oscillate sinusoidally and if an electron is initially at $\mathbf{k}_0$, (72) shows that subsequently

$$\left. \begin{aligned} k_x &= k_{0x} \cos \omega_c t + (m_x/m_y)^{1/2} k_{0y} \sin \omega_c t \\ k_y &= k_{0y} \cos \omega_c t - (m_y/m_x)^{1/2} k_{0x} \sin \omega_c t. \end{aligned} \right\} \tag{1.73}$$

and

An impulsive electric field in the x-direction changes k_{0x} and v_{0x} for each electron by the same amount, leaving k_{0y} and v_{0y} unchanged. Thus the development of $\mathbf{J}(t)$ follows that of $\mathbf{k}$ in (73) with the terms in k_{0y} omitted:

$$J_x = (ne^2/m_x) \cos \omega_c t \cdot e^{-t/\tau} \tag{1.74}$$

and

$$J_y = -(ne^2/m_x^{1/2} m_y^{1/2}) \sin \omega_c t \cdot e^{-t/\tau}. \tag{1.75}$$

The coefficients are supplied to lead to the correct initial response to unit impulse, and exponential decay is inserted to describe the development of the impulse response with constant relaxation time. Integrating over t gives the response to a steady field of unit strength, that is

$$\sigma_{xx} = ne^2\tau/m_x(1 + \omega_c^2\tau^2)$$

and

$$\sigma_{yx} = -ne^2\omega_c\tau^2/(m_xm_y)^{1/2}(1 + \omega_c^2\tau^2).$$

Similarly

$$\sigma_{yy} = ne^2\tau/m_y(1 + \omega_c^2\tau^2).$$

Inserting these expressions into (14) we see that $\rho_{xx} = m_x/ne^2\tau$ and $\rho_{yy} = m_y/ne^2\tau$ at all values of B.

Anisotropy of the energy surfaces is not in itself sufficient to cause the appearance of magnetoresistance although, as the following rather extreme example will show, it may be responsible for a mild and saturating effect. Up to this point only equality or near equality of electron and hole concentrations has provided a mechanism for any striking magnetoresistance; but there is another, open orbits resulting from Fermi surfaces that are not closed, which will be discussed fully in chapter 3.

99

Square Fermi surface with constant relaxation time

Artificial as this model may seem, it possesses features which illustrate to a somewhat exaggerated degree the properties of some real metals, and we shall have occasion to refer to it on one or two occasions later. It is therefore worth spending a little time on it, even to the extent of analysing its behaviour by two different methods, equivalent in effect but serving to exemplify different techniques that are available. Neither method employs the effective path, though this too could be deployed if it were simpler, which here it is not.

Method 1: The square Fermi surface is shown in fig. 18. As we are concerned only with the form of the variation of $\rho(B)$ we shall not trouble to evaluate constants. Thus we suppose B to drive electrons round the square at constant speed V, so that the time for a revolution is $4K/V$ and $\gamma \equiv \omega_c\tau = \frac{1}{2}\pi V\tau/K$. Also we suppose that an electric field $\mathscr{E}_n$, lying in the plane of the paper and normal to one of the sides, creates or destroys particles at a rate $a\mathscr{E}_n$ per unit length. The rate of increase of N, the excess number per unit length, is then given by the equation of continuity,

$$\dot{N} = a\mathscr{E}_n - N/\tau - V\,\partial N/\partial s, \tag{1.76}$$

s being measured anticlockwise round the side of the square. In the steady

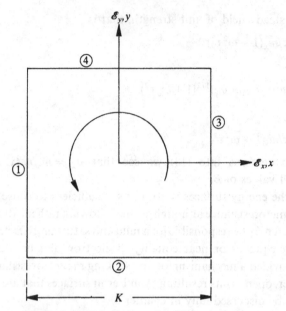

Figure 1.18 Calculation of transverse magnetoresistance for a square Fermi surface; **B**, normal to the plane, causes anticlockwise circulation.

state $\dot{N} = 0$ and the general solution is

$$N = a\mathscr{E}_n\tau + Ae^{-s/V\tau}, \tag{1.77}$$

where A is an arbitrary constant.

We proceed to calculate ρ_{xx} directly, allowing $\mathscr{E}_x$ and $\mathscr{E}_y$ to be present in such proportion that J_y vanishes. On side 1, where $\mathscr{E}_x$ creates electrons, (77) takes the form

$$N_1 = a\mathscr{E}_x\tau + A_xe^{y/V\tau}, \tag{1.78}$$

and on side 3, where electrons are destroyed,

$$N_3 = -a\mathscr{E}_x\tau - A_xe^{-y/V\tau}. \tag{1.79}$$

Similarly,

$$N_2 = a\mathscr{E}_y\tau + A_ye^{-x/V\tau} \tag{1.80}$$

and

$$N_4 = -a\mathscr{E}_y\tau - A_ye^{x/V\tau}. \tag{1.81}$$

Now if $J_y = 0$, there are no excess electron on sides 2 or 4, and $\int_{-K/2}^{K/2} N_2 dx = 0$. Hence,

$$A_y = -(aK\mathscr{E}_y/2V)\operatorname{cosech}(\pi/4\gamma) \tag{1.82}$$

and N_2 and N_4 are found in terms of $\mathscr{E}_y$. To complete the calculation we match $N_1(y = \frac{1}{2}K)$ with $N_4(x = -\frac{1}{2}K)$ at the upper left-hand corner, and $N_3(y = \frac{1}{2}K)$ with $N_4(x = \frac{1}{2}K)$ at the upper right-hand corner. A little manipulation yields

$$A_x = -2(\sinh D + D \cosh D) a\tau\mathscr{E}_x/(\sinh 2D + 2D \cosh 2D),$$

$$(1.83)$$

where $D = \pi/4\gamma$. Finally we form J_x as proportional to $\int_{-K/2}^{K/2} N_1 \mathrm{d}y$, and express ρ_{xx} as $\mathscr{E}_x/J_x$; when the constant multiplier is eliminated by referring everything to the zero-field resistivity, we have

$$\rho_{xx}/\rho_0 = D(2D \cosh 2D - \sinh 2D)/$$
$$[(2D^2 + 1)\cosh 2D - 2D \sinh 2D - 1].$$

$$(1.84)$$

For the high-field behaviour, when $D \to 0$, expansion of the terms shows that $\Delta\rho_{xx}(\infty)/\rho_0 = \frac{1}{3}$, confirming that the shape of a closed Fermi surface is not a very important influence on magnetoresistance.

The detailed behaviour, shown in fig. 19, exhibits a point of interest in that there is no quadratic term at low values of $\omega_c\tau$. This is a consequence of the sharp corners on the Fermi surface, which cause (84) to be non-analytic as $1/D \to 0$. However small $\omega_c\tau$ may be there are electrons near the corners, in number proportional to $|\omega_c\tau|$, which are turned through $\pi/2$ before suffering collision. When something like this appears in practice, as in figs. 5 and 6, one may infer, if not perfectly sharp corners, at least regions of high curvature on the Fermi surface. Without geometrically sharp corners there

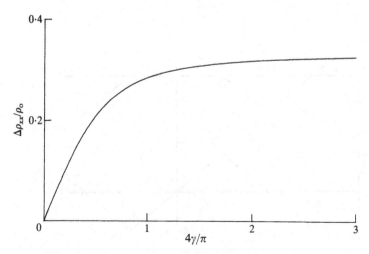

Figure 1.19 Magnetoresistance of the square Fermi surface of fig. 18, according to (84); $4\gamma/\pi = 1/D = 2V\tau/K$.

is always an initial quadratic form which may be followed by a nearly linear range at values of $\omega_c \tau$ much less than unity.

Method 2: Instead of deriving ρ_{xx} directly we now proceed by way of $\sigma_{xx} = \sigma_{yy}$ and $\sigma_{xy} = -\sigma_{yx}$, by following the impulse response of the current as a whole; the simple shape of the Fermi surface makes it unnecessary to take each element separately. Unit impulsive $\mathscr{E}_x$ creates new particles on side 1 and destroys as many on side 3, leaving the other sides unchanged. Subsequent steady motion round the square at speed V causes the current, initially J_{x0}, to oscillate in the absence of collisions as in fig. 20(a), while J_y grows from zero and oscillates similarly, in phase quadrature, as in fig. 20(b). We now let each decay as $e^{-t/\tau}$ and integrate over time to obtain σ_{xx} and σ_{yx}.

In the first half-cycle, $0 < t < \pi/\omega_c$, $J_x = J_{x0}(1 - 2\omega_c t/\pi)e^{-t/\tau}$, and

$$\int_0^{\pi/\omega_c} J_x \, dt = J_{x0}\tau[1 + e^{-4D} - (1 - e^{-4D})/2D]. \tag{1.85}$$

Subsequent half-cycles make similar contributions, successively multiplied by $-e^{-4D}$, to give a geometrical progression whose sum is $(1 + e^{-4D})^{-1}$. When this factor is attached to (85) the resulting σ_{xx} is simply expressed:

$$\sigma_{xx} = J_{x0}\tau(1 - \tanh 2D/2D). \tag{1.86}$$

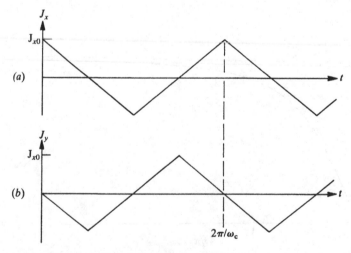

Figure 1.20 Impulse response of J_x and J_y for the square Fermi surface of fig. 18.

A similar calculation gives

$$\sigma_{yx} = J_{x0}\tau(1 - \operatorname{sech} 2D)/2D. \tag{1.87}$$

On forming $\rho_{xx} = \sigma_{yy} = \sigma_{xx}/(\sigma_{xx}^2 + \sigma_{yx}^2)$, (84) is recovered without difficulty.

It is worth noting that as $\omega_c\tau \to \infty$, σ_{yx} is found by summing the area of triangles which alternate in sign as they diminish slowly, and (as is well-known from elementary Fraunhofer diffraction theory) the sum is half the area of the first term, and is insensitive to the details of the decrement. This tallies with the principles underlying the derivation of (51). On the other hand the curve in fig. 20(*a*) starts with a half-triangle and application of the simple summation rule would lead to zero. The difference is obvious in the series expansion of (86) and (87) in terms of D:

$$\sigma_{xx} \sim J_{x0}\tau \times \tfrac{4}{3}D^2, \quad \sigma_{yx} \sim J_{x0}\tau \times D.$$

When one is forced to go to second order in the expansion, as for σ_{xx}, the details of the decay, i.e. the precise variation of scattering round the Fermi surface, must be included in the calculation. This will be made more explicit in chapter 3.

1

2 Measurement

It is only too easy to assume that the measurement of magnetoresistance is an ancient art that offers no challenge to an experimenter equipped with the finest modern instrumentation. One should not forget, however, that the largely outmoded moving-coil galvanometer was as nearly perfect a device as can be made, in that its sensitivity was limited only by Brownian motion and Johnson noise appropriate to a room-temperature source.[1] If this sensitivity was rarely available in practice it was because of other noise sources in the equipment, such as would not be eliminated by replacing the galvanometer with a (probably) more noisy digital voltmeter.

To appreciate the problem let us consider a conventional measurement of the magnetoresistance of potassium in the form of a 'matchstick' sample[2] – a rectangular bar of 2 mm square cross-section and 2–3 cm long; anything thinner is hard to handle without damage. To be sure, potassium is one of the more difficult materials, but it has been studied extensively, as we shall learn, and serves as a realistic example. To measure its resistance electrodes are attached at the ends to lead current in and out, and two potential electrodes are placed about 2 cm apart on the same long face. A resistance ratio r_0 of 5000, typical of a good sample, produces a resistance of $6 \times 10^{-8} \, \Omega$ between the potential electrodes. If one is to measure to 1% a sensitivity of 10^{-9} V is needed even when a current of 2 A is passed; as fig. 1.3 shows, 1% is by no means an excessive requirement. Under ideal conditions a DC voltage of 10^{-9} V presents a moderately severe test of experimental design, but conditions in a low-temperature magnetoresistance measurement are not ideal for two principal reasons – the presence of a strong magnetic field, and thermoelectric effects in the leads connecting the potential contacts at, say, 4 K to the voltmeter at room temperature.* If one was so foolish as to use

---

* SQUIDs,[3] or other superconducting devices, allow the voltage measurement to be made at the temperature of the sample, thus eliminating thermoelectric disturbances as well as enhancing the voltage sensitivity very greatly; but they have to be shielded from the applied field, and this restricts their usefulness.

different metals for the two potential leads, there might well be a thermoelectric voltage amounting to 10^{-3} V; even apparently identical wires can turn out to generate annoying drifts of voltage as temperatures in the cryostat fluctuate. Troubles caused by the magnetic field are less of a worry now, with superconducting magnets, than formerly when fluctuations in the current exciting an electromagnet induced spurious EMFs in the measuring circuit. Even now, when a superconducting magnet carries a very steady persistent current, vibrations of the sample and leads can still induce troublesome EMFs; for a field of 5 T threading the measuring circuit induces 10^{-9} V if the area of the circuit oscillates with an amplitude of only 3×10^{-5} mm^2 at a frequency of 1 Hz.

When one also takes into account the hazard of a Hall field that may be 200 times stronger than the longitudinal field one begins to appreciate the need for sceptical appraisal of any published results, especially those that present only the automated output or digested analysis with no account of the method used or precautions adopted. The following discussion of a variety of alternative techniques does not go into such detail as to provide a handbook for the novice. Rather it is a guide to the literature and, more importantly, a critical survey whose final conclusion is that no method, however elegant or carefully thought through, is proof against the insensate malice of nature that every experimenter, even in a traditional area, ought to expect as a matter of course.

Since we are concerned as much with what can go wrong as with what ought to go right, we have to examine the effects of departures from ideality, whether they are imperfections of sample shape, misalignments in the magnetic field, poor electrode attachment, or any others. This involves solution of the field equations with uncompromising boundary conditions; that is, the sort of problem to which the only general solutions are hopelessly uninformative. For this reason we shall look at model problems that reveal the nature of the effects to be expected, and use the insights so gained to discuss specific experiments qualitatively. Let us start by formulating the field equations.

Field equations

In the steady state $\mathbf{B} = 0$ and there is no electromagnetically induced electric field. Consequently curl $\mathscr{E} = 0$, and because any space charges are unchanging, div $\mathbf{J} = 0$. When the material is uniform on the scale of the mean free path $\mathbf{J}$ and $\mathscr{E}$ are locally related by $J_i = \sigma_{ij}\mathscr{E}_j$, and these statements suffice to define the field equations. It may be noted that the polarizability of the atom cores does not enter the problem – there is no place for a displacement vector $\mathbf{D}$; if this polarizability happens to cause conflict with the matching of $\mathbf{J}$ and $\mathscr{E}$, conduction of charge will establish a compensating space charge. Thus at an interface between different materials continuity of the normal component of $\mathbf{J}$ will always override continuity of the normal component of $\mathbf{D}$, since a surface charge suffices to destroy the latter.

The irrotational property of $\mathscr{E}$ allows V to be defined, as usual, so that $-\operatorname{grad} V = \mathscr{E}$. Then,

$$J_i = -\sigma_{ij}\partial V/\partial x_j$$

and

$$-\operatorname{div}\mathbf{J} = \sigma_{ij}\partial^2 V/\partial x_i\partial x_j = 0. \tag{2.1}$$

In general σ_{ij} may be dissected into a symmetrical and an antisymmetrical part, and Cartesian axes may be chosen so that the symmetrical part is diagonalized (if $\mathbf{B}$ does not coincide with an axis of crystal symmetry the required axes for diagonalization will not coincide with the crystal axes, but we shall not discuss any case where this needs detailed consideration). When this choice of axes is made, the off-diagonal terms in σ_{ij} are antisymmetrical and disappear from (1) in the summation over repeated indices. We then have

$$\sigma_{xx}\partial^2 V/\partial x^2 + \sigma_{yy}\partial^2 V/\partial y^2 + \sigma_{zz}\partial^2 V/\partial z^2 = 0, \tag{2.2}$$

which can be reduced to the familiar form of Laplace's equation by scaling the axes. In terms of new orthogonal axes ξ, η, ζ such that $\sigma_{xx}^{1/2}\xi = cx$, $\sigma_{yy}^{1/2}\eta = cy$, $\sigma_{zz}^{1/2}\zeta = cz$, c being an arbitrary constant, $\nabla^2 V = 0$.

Suppose we wish to find the current pattern in a uniform block with current leads attached at two points. We first scale the dimensions of the block by factors $\sigma_{xx}^{1/2}$ etc., possibly changing its shape considerably in the process. Then by standard procedures, computation or experiment, we solve Laplace's equation for the new block and draw equipotential surfaces compatible with the electrode positions. The scaling is next reversed, the equipotential surfaces being carried with the material of the block as it regains its original shape. From then on $\mathscr{E}$ follows as $-\operatorname{grad} V$ and J_i as $\sigma_{ij}\mathscr{E}_j$. In the scaled model, $\mathscr{E}$ as derived from V in non-divergent, but in the actual block $\operatorname{div}\mathscr{E}$ need not vanish, and space charges are to be expected, though we rarely have occasion to determine their distribution. One must, however, be aware of the possibility when attempting to sketch field and current lines.

Current jetting

Let us apply these principles to an isotropic material, with **B** lying along the z-axis, so that σ_{ij} has the form (1.18). Then the scaled coordinates may be chosen with $\xi = x, \eta = y$ and $\zeta = (\sigma_1/\sigma_3)^{1/2}z$. In a pure sample, with $\omega_c\tau \gg 1, \zeta$ is roughly $z/\omega_c\tau$, and the sample shape may need to be very strongly foreshortened along **B** to construct the scaled model.

The simplest problem involves a point contact injecting current into a semi-infinite sample whose plane surface is normal to **B**, as in fig. 1. In the scaled model (a) the equipotentials are spherical, according to

$$V = M + N/r, \tag{2.3}$$

where r is the radius vector measured from the injection point, and M and N are constants. When the scaling is reversed (b) x and y are unchanged while the z-coordinate is pulled out by a factor $R = (\sigma_3/\sigma_1)^{1/2}$, so that in the real sample,

$$V = M + N/(r_t^2 + z^2/R^2)^{1/2}, \tag{2.4}$$

r_t being the transverse component of the radius vector. Then,

$$\left.\begin{aligned}
\mathscr{E}_{\parallel} &= -\partial V/\partial z = (Nz/R^2)/(r_t^2 + z^2/R^2)^{3/2}, \\
\mathscr{E}_{\perp} &= -\partial V/\partial r_t = (r_t R^2/z)\mathscr{E}_{\parallel}.
\end{aligned}\right\} \tag{2.5}$$

and

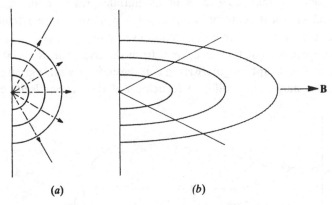

(a) (b)

Figure 2.1 Scaling to restore Laplace's equation when σ_{ij} is anisotropic. In (a) current injected into an isotropic conductor spreads evenly, with half lying within the $60°$ cone indicated by the extended lines. In (b) the diagram is redrawn with the semicircular equipotentials drawn out into ellipses, in this case with $R = 3.1$; half the current now lies in a $28°$ cone.

Ignoring for the moment the Hall conductivity, we put

$$J_\parallel = \sigma_3 \mathscr{E}_\parallel = R^2 \sigma_1 \mathscr{E}_\parallel = N\sigma_1 z/(r_t^2 + z^2/R^2)^{3/2}, \qquad (2.6)$$

and

$$J_\perp = \sigma_1 \mathscr{E}_\perp = (r_t/z)J_\parallel. \qquad (2.7)$$

According to (7), and not unexpectedly, **J** without the Hall current would diverge radially from the injection point, but (6) shows that it is not uniformly spread through the hemisphere. Integrating (6) leads to the result that the fraction of current lying within θ from the normal is $1 - (1 + R^2 \tan^2 \theta)^{-1/2}$; half the current lies within a cone of semi-angle $\tan^{-1}(\sqrt{3}/R)$, e.g. $\frac{1}{2}°$ for a sample of potassium having $R \sim \omega_c \tau = 200$.

The powerful preference for current to flow along **B** when $\sigma_{zz} \gg \sigma_{xx}$ is called *current jetting*[4] and must not be overlooked in the design of experiments to measure magnetoresistance. One must also remember that this is not the whole story, for $\sigma_{xy} (= \sigma_2)$ is likely to be considerably larger than σ_1 and will result in a strong circulating component of current. The lines of current flow lie on cones as just described, but form helices like that shown in fig. 2. When $\sigma_3 \gg \sigma_2 \gg \sigma_1$ the circulating current is dominated by $\sigma_2 \mathscr{E}_\perp$ and the longitudinal current by $\sigma_3 \mathscr{E}_\parallel$, so that the lines make an angle of approximately $\tan^{-1}(\sigma_2 \mathscr{E}_\perp/\sigma_3 \mathscr{E}_\parallel)$ with **B**. If $\sigma_3 = R\sigma_2$, (5) shows that this angle is $\tan^{-1}(r_t R/z)$, or $\pi/4$ on a typical cone in the jet for which $r_t/z = 1/R$.

Current jetting creates a serious problem in measurements of the longitudinal magnetoresistance in such materials as potassium, or any other where σ_{zz} may be orders of magnitude greater than σ_{xx}. In a matchstick sample the current injection must be uniform enough to allow it to have covered the whole cross-section evenly by the time it passes the potential electrodes. If these are 5 mm from the ends, and $R = 200$, there must be no significant irregularities in **J** at the ends on a scale greater than 25 μm. This may not be difficult to achieve if the current leads can be

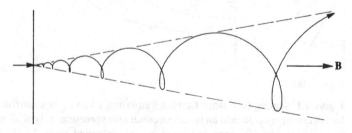

Figure 2.2 A line of current flow for point injection when $R = 10$. The conical helix is viewed along a line lying at $72°$ to **B**.

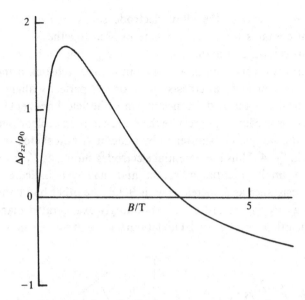

Figure 2.3 Longitudinal magnetoresistance of an antimony crystal at 1.5 K showing strong negative magnetoresistance, probably an artefact of current jetting.

soldered on, but with potassium and other hardly wettable metals it is easy to make a serious mistake. The current may be injected only near the axis of the sample, and fail to spread to the surface, so that the potential leads remain unaware of its existence. In this way a negative resistance may be recorded, as in fig. 3 (whose origin I suppress to avoid needless embarrassment to the author). Ueda and Kino[5] have studied the effect carefully, and give many references.

Even if soldering is feasible it is still worth taking precautions to ensure as uniform a current density as possible, by attaching to each end of the matchstick a rod of the same cross-section and made of a resistive alloy. Minor variations in contact resistance will then have minimal effect since rearrangement of the current flow in the alloy will be costly in terms of extra dissipation (the current distributes itself so as to give the least dissipation).

As for the problem presented by potassium and other reactive metals, the technique adopted by Lass[6] is usually effective. He found that a bead of mercury placed on the end of the matchstick would amalgamate with the potassium if the end was scratched with a needle under the mercury surface; and a previously amalgamated alloy and rod could then be firmly attached by freezing. Fletcher[7] further improved the process by machining a fine pyramidal grid on the end of the alloy electrode so as to produce a multitude of point contacts when the two were pressed together.

This is not, however, the end of the story of current jetting, for it can be responsible for serious errors even in a well-connected matchstick if the sample is bent, or of non-uniform cross-section, or not perfectly aligned with the magnetic field; or if the field is non-uniform or the field-lines bent.[4] No detailed analysis is needed to appreciate the nature of the problem; one has only to draw the sample foreshortened by a factor R and sketch the equipotentials, as in fig. 4. Thus the potential electrodes shown as V_1 in (c) will record only a small potential difference and may well indicate a negative magnetoresistance as R increases with B. On the other hand, the electrodes shown as V_2 will record a correspondingly exaggerated magnetoresistance. The other examples need no labouring, but it may be noted

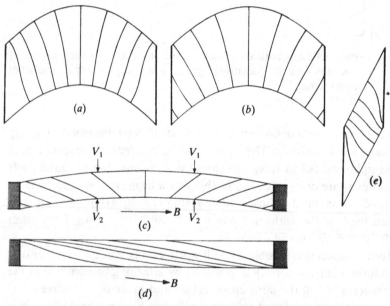

Figure 2.4 Effects of specimen distortion etc. in longitudinal magnetoresistance, as revealed by evenly spaced equipotentials. (a) Solution of Laplace's equation in a bent strip, with the left and right edges as equipotentials, and no current flowing across top and bottom edges. (b) the same except that current crosses the vertical edges from highly resistive electrodes. If $\omega_c\tau = 5$, a real strip (c) would scale to (b) and have equipotentials as shown; potential electrodes $V_1 V_1$ and $V_2 V_2$ would give readings differing by more than a factor of 2. The undistorted strip (d) scales to (e) when $\omega_c\tau = 20$ and **B** is not quite parallel; the resulting deformation of the equipotentials in (d) is very marked.

that a uniformly tapered sample need not suffer great error if the electrodes are resistive enough to ensure an even spread of current at both ends and straight lines of current flow. Surface imperfections,[8] however, in the form of scratches or pits between the potential electrodes may deflect the current away from the surface sufficiently to disturb the answer.

The moral is that measuring longitudinal magnetoresistance with matchstick samples is by no means the straightforward matter that one is tempted to assume at first. Anyone reporting results obtained with materials of high $\omega_c\tau$ should describe his procedure and his checks in some detail if he is to carry conviction.

Matchstick sample in a transverse field

When **B** lies normal to a matchstick sample the scaling process just described allows us to see the potential as obeying Laplace's equation in a sample reduced in thickness, along **B**, by a factor R. When $R \gg 1$ it may well be permissible to treat the sample as a thin foil and replace it by a two-dimensional model. In effect, the conductivity along **B** is so high that V does not change from one side to the other. The problem of attaching current leads is now, if anything, even more acute than with **B** longitudinal, for even if they are well soldered the difference between their conductivity and that of the sample can produce a highly non-uniform current distribution. On the other hand, it is not so hard for the current to spread to fill the cross-section.

Fig. 5 illustrates how the origin of the difficulty lies in the Hall effect. A uniform current in resistive electrodes ($\omega_c\tau \ll 1$) sets up equipotentials normal to the current flow (*a*); but in a sample showing the Hall effect the equipotentials lie at the Hall angle to the normal (*b*). Since the Hall angle may be close to $\pi/2$ when $\omega_c\tau \gg 1$, they may lie very nearly along the sample

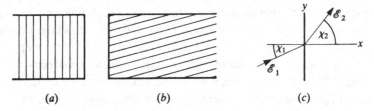

      (*a*)   (*b*)   (*c*)

Figure 2.5 Equipotentials in a current-carrying strip with **B** normal to the page, (*a*) when the Hall angle is small, as in a resistive material, and (*b*) when the Hall angle is 75° ($\omega_c\tau \sim 3.7$). (*c*) illustrates the boundary condition for $\mathscr{E}$ at an interface between materials of different Hall angle.

axis. At the junction V is continuous, and clearly considerable adjustment of the current pattern may be required to meet this condition. In fact the current is concentrated at one side of the junction.

A general solution is probably not possible in closed form but we can make good progress with soluble models. Let us assume the materials obey (1.18), being uniform and isotropic in the plane of the model foil. Then it is easily shown that the vanishing of curl $\mathscr{E}$ and div $\mathbf{J}$ imply curl $\mathbf{J} = 0$; hence a velocity potential ϕ may be defined such that $\mathbf{J} = -\operatorname{grad}\phi$ and $\nabla^2\phi = 0$. At all points $\mathscr{E}(= -\operatorname{grad} V)$ and $\mathbf{J}$ lie at the Hall angle to one another, and if we solve for one ($\mathbf{J}$, say) we can construct the other without trouble.

At an interface between two media (fig. 5c) $\mathscr{E}_y$ and J_x are continuous. Now $\mathscr{E}_y = -\rho_2 J_x + \rho_1 J_1$, so that

$$-\rho_2^{(1)}J_x + \rho_1^{(1)}J_y^{(1)} = -\rho_2^{(2)}J_x + \rho_1^{(2)}J_y^{(2)},$$

and therefore

$$-\rho_2^{(1)} + \rho_1^{(1)}\tan\chi_1 = -\rho_2^{(2)} + \rho_1^{(2)}\tan\chi_2. \tag{2.8}$$

If a pattern of current flow in one medium is governed by a potential that satisfies Laplace's equation, the mirror image of this pattern in the other medium will also do so and will automatically ensure current continuity across the interface. Then continuity of $\mathscr{E}_y$ is ensured by choosing $\chi_1 = -\chi_2 = \chi$, say, to satisfy (8); i.e.

$$\tan\chi = (\rho_2^{(1)} - \rho_2^{(2)})/(\rho_1^{(1)} + \rho_1^{(2)}). \tag{2.9}$$

The current distribution in the sample and the electrode is determined by the requirement that $\mathbf{J}$ must not cross the sides and must cross the interface at a constant angle χ. This awkward set of boundary conditions inevitably leads to singular behaviour at the sides of the interface, where $\mathbf{J}$ changes direction abruptly.

The nature of the singularity can be found by assuming, as is plausible, that close to one side the other side has no noticeable influence on the flow. The solution can now be derived from that of a standard problem in hydrodynamics, the irrotational flow of an inviscid fluid round a sharp corner,[9] or the corresponding electrostatic problem, a charged conducting wedge.[10] In the latter case, if we use polar coordinates with the corner of the wedge as origin, the potential outside the wedge is given by

$$V = r^a \sin(a\theta). \tag{2.10}$$

When $V = 0$, $\theta = 0$ or π/a for all r, and this defines the external angle of the wedge as π/a. In fig. 6 (which is a mirror image of the conventional

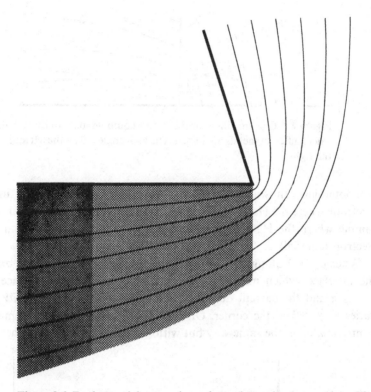

Figure 2.6 Equipotentials round a charged conducting wedge. The shaded region is used for the current lines in fig. 7.

coordinate system) $a = 0.625$, so that the internal angle of the wedge is 72°. Because the wedge is infinite there is no intrinsic length scale in the problem and all equipotentials are similar in shape, as follows immediately from (2.10); along any line of constant θ, $V \propto r^a$ and is singular at the origin if $a < 1$, i.e. a salient corner.

A curious property of V as defined by (10) is that not only do all equipotentials cut a radius, $\theta =$ constant, at the same angle, which is a consequence of their similarity, but this angle is $a\theta$. The corollary of this simple proportionality is that if we superpose on the picture of the equipotentials the same picture turned about the origin through an angle θ_0, the two sets of lines everywhere intersect at the same angle, $a\theta_0$. Thus the solution may be used to draw both sets of equipotentials (V and ϕ) and both sets of field lines ($\mathscr{E}$ and $\mathbf{J}$) whatever the Hall angle. In this way the current lines in fig. 7 have been constructed, taking the shaded area of fig. 6 to describe $\mathbf{J}$ in the electrode and its mirror image for $\mathbf{J}$ in the sample; the

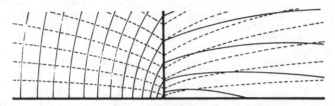

Figure 2.7 Current lines (broken) and equipotentials (full) at a corner where different metals are joined. The Hall angle is 0 on the left and 80° on the right.

equipotentials, shown as full lines, are drawn normal to **J** in the electrode where the Hall angle is assumed to be zero, and at 10° to **J** in the sample where the Hall angle is assumed to be 80° ($\omega_c\tau = 5.7$ in a free-electron material).

When $\theta = \pi/2$, along the interface, the lines of flow are inclined at $a\pi/2$ to the interface, which must be $\pi/2 - \chi$, χ being given by (9). Hence $a = 1 - 2\chi/\pi$ and the current density, J_x, which is proportional to $(\partial V/\partial r)_\theta$, varies as $y^{-2\chi/\pi}$ at the corner. On the opposite side the solution may be approximated in the same way but with $-\chi$ replacing χ, so that here the

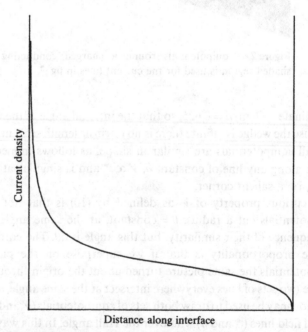

Figure 2.8 Suggested current distribution across the interface between strips with Hall angle 0 and 80°.

current density varies as $y^{2\chi/\pi}$. These are, of course, only limits of the true variation and no such elementary analysis will give an expression that holds right across the interface. However, by matching the limiting forms, both in amplitude and gradient, at the middle of the interface one obtains a current distribution as in fig. 8 which is probably a very fair representation.

The non-uniformity of current is minimized by making χ as small as possible. Since with high $\omega_c\tau$ in the sample, and resistive electrodes, the dominant resistivities in (9) are $\rho_2^{(2)}$ and $\rho_1^{(1)}$, the electrode resistivity should overwhelm the highest Hall resistivity in the sample, B/ne for a free electron metal; this is usually possible. In any case, however uneven the current distribution at the interface, it evens itself out within one or two diameters of the matchstick – provided (1.18) fairly represents the conductivity, there is no jetting problem in a transverse field to cause trouble as in a longitudinal field.

It is worth remarking in passing that when the Hall effect and transverse magnetoresistance are measured in semiconductors, with metallic electrodes, $\rho_2^{(2)}$ and $\rho_1^{(2)}$ are usually very much greater than $\rho_1^{(1)}$, and (9) shows that the current crosses the interface at the Hall angle to the axis, which may be nearly $\pi/2$. If it is $\pi/2 - \varepsilon$, the current density near one corner varies as $y^{-1+\varepsilon}$, only escaping divergence because of ε; in other words, most of the current enters the sample very near the corner.[11] But this is not our worry.

The proviso that the analysis up to this point assumes the validity of (1.18) is meant seriously, for real metals may depart very markedly from this ideal, as we shall find out in the next chapter. It is possible for σ_{ij} to be extremely anisotropic in the plane transverse to **B**, with principal axes along which σ_{xx} may be orders of magnitude greater than σ_{yy}. One must then apply the same scaling process as we found useful for longitudinal magnetoresistance, contracting the x-dimension relative to the y-dimension by a factor $(\sigma_{xx}/\sigma_{yy})^{1/2}$, in order to construct a model in which Laplace's equation holds. If the principal axis of easy conduction does not lie along the length of the matchstick, current jetting along this axis may confine the current to a narrow channel that largely avoids the potential electrodes. The resulting negative magnetoresistance can be so great as to mislead experimenters into the belief that they have discovered a transition to superconductivity in high magnetic fields.[12]

Sample imperfection – inhomogeneity

It is possible for relatively minor inhomogeneities in the sample to influence the current flow considerably. This was pointed out by Herring[13] in

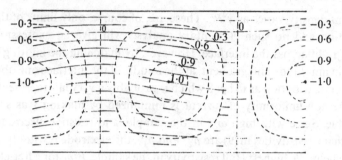

Figure 2.9 Current lines (full) and contours of fluctuations in ρ_2 (broken) that give rise to the non-uniform current. The labels against the contours are to be multiplied by ρ_1/ρ_2 $(1/\omega_c\tau$ for a free electron metal) to give $\Delta\rho_2/\rho_2$.

connection with semiconductors which are peculiarly sensitive to their level of doping and consequently liable to inhomogeneity if the doping is not strictly uniform. Metals are probably less seriously affected, but the possibility must be borne in mind, and a model calculation will serve to illustrate the point. Since Herring's analysis makes clear that variations of σ_{xy} are much more serious than variations of σ_{xx}, we shall consider a metal obeying (1.18), with σ_1 uniform and σ_2 non-uniform.

We shall synthesize a solution, starting with an assumed non-uniform current and finding what variations of σ_2 are needed to produce it. Consider a two-dimensional strip of width $2a$ in which $\mathbf{J}$ pursues a wavy course, as in fig. 9:

$$\left.\begin{aligned} J_x &= J_0 + J_1 \sin kx \sin (\pi y/2a) \\ J_y &= (2ka/\pi)J_1 \cos kx \cos (\pi y/2a). \end{aligned}\right\} \tag{2.11}$$

This current is non-divergent and is accompanied by an electric field:

$$\left.\begin{aligned} \mathscr{E}_x &= \rho_1[J_0 + J_1 \sin kx \sin (\pi y/2a)] \\ &\quad + \rho_2(2ka/\pi)J_1 \cos kx \cos (\pi y/2a) \\ \mathscr{E}_y &= - \rho_2[J_0 + J_1 \sin kx \sin (\pi y/2a)] \\ &\quad + \rho_1(2ka/\pi)J_1 \cos kx \cos (\pi y/2a). \end{aligned}\right\} \tag{2.12}$$

The field is irrotational, i.e. $\partial\mathscr{E}_x/\partial y = \partial\mathscr{E}_y/\partial x$, and applying this condition to (2.12) gives an equation for the variation of ρ_2:

$$\rho_1 J_1(\beta^2 + k^2)\sin \xi \cos \eta + k^2 J_1 \cos \xi \cos \eta(\partial\rho_2/\partial\eta)$$
$$+ \beta^2[J_0 + J_1 \sin \xi \sin \eta](\partial\rho_2/\partial\xi) = 0. \tag{2.13}$$

in which $\beta = \pi/2a$, $\xi = kx$ and $\eta = \beta y$. If the waviness is small and $J_1 \ll J_0$, we need retain only the dominant terms of (13):

$$\rho_1 J_1 (\beta^2 + k^2) \sin \xi \cos \eta + \beta^2 J_0 \partial \rho_2 / \partial \xi \approx 0,$$

so that

$$\rho_2 \approx \bar{\rho}_2 [1 + (\rho_1 J_1 / \rho_2 J_0)(1 + k^2/\beta^2) \cos \xi \cos \eta]. \qquad (2.14)$$

If k and β are of similar magnitude the relative variations of ρ_2 from its mean $\bar{\rho}_2$ are largely determined by $(\rho_1 J_1 / \rho_2 J_0)$, which is about $\omega_c \tau$ times smaller than the current variation, J_1 / J_0. That is to say, we must expect variations in the Hall constant to be reflected in current variations amplified by a factor $\omega_c \tau$. The contours in fig. 9 are drawn for $J_1 = J_0/5$; if $\omega_c \tau = 200$, ρ_2 has to vary only by $\pm 0.2\%$.

This example is sufficient to indicate the need for homogeneity, but it is no more than an illustration. Often enough the scale of inhomogeneity will be much smaller than the width of the sample, and it may then turn out that the two-dimensional model is inadequate; at a small irregularity some of the current flow may be deflected into the direction of **B**. We shall return in chapter 5 to more general considerations involving inhomogeneous samples.

Voids

Since many gases are able to dissolve in liquid metals but not in the solid phase it is possible for a sample to contain small gaseous inclusions which are sometimes blamed for anomalous magnetoresistive behaviour, especially in potassium.[14] Let us therefore see how the measured resistance of a sample obeying (1.18) is affected by a void, assumed spherical.[15] The first thing to do is to apply such scaling factors as will make the potential obey Laplace's equation. The sphere becomes an oblate spheroid, if $\sigma_3 > \sigma_1$, with axes in the ratio $1:1:R^{-1}$, and we must find a pattern for $\mathscr{E}$ that generates no current across the surface of the void. The problem is of the same type as the well-known problem of an ellipsoidal dielectric inclusion in a different dielectric matrix, both dielectrics being isotropic.[16] When a uniform field is applied the inclusion is polarized uniformly, though not necessarily in the same direction as the applied field. The same analysis applies to a conducting ellipsoidal inclusion in a matrix of different conductivity, and it has been shown[17] that the same general principle holds when both matrix and inclusion (which must be ellipsoidal) have arbitrary anisotropic conductivity tensors – that is, the internal field is uniform and therefore the current density is also uniform in the inclusion.

It is convenient to replace the polarized inclusion by a surface charge distribution within a uniform matrix. Thus instead of the void we insert charges on a spheroidal shell that simulate uniform polarization $\mathbf{P}$ – the charge density is $\mathbf{n} \cdot \mathbf{P}$ at a point on the shell where the surface normal is represented by unit vector $\mathbf{n}$. According to the theory of polarized ellipsoids, if $\mathbf{P}$ lies along a principal axis the field generated within the shell is $-L\mathbf{P}$, where L is the depolarizing factor (0 for a rod parallel to $\mathbf{P}$, $1/\varepsilon_0$ for a slab normal to $\mathbf{P}$, $1/3\varepsilon_0$ for a sphere, etc.). By assuming this result, and quoting appropriate values for L, we bypass the lengthiest part of the calculation. All that is necessary is to choose $\mathbf{P}$ so that the normal component of $\mathscr{E}$ just outside the void vanishes at a few selected points; relying on the general theory to ensure that it then vanishes everywhere.

To illustrate the procedure, take first the influence of a void in a measurement of longitudinal magnetoresistance. Here the applied electric field is parallel to $\mathbf{B}$ and the sphere has to be foreshortened by a factor R along the same direction. Symmetry assures us that $\mathbf{P}$ will also lie in the same direction. Then at the pole of the spheroid, where the applied field $\mathscr{E}_0$ emerges, the internal field is $\mathscr{E}_0 - L\mathbf{P}$, the charge density is $\mathbf{P}$, and therefore, by Gauss' theorem, the external field is $\mathscr{E}_0 + (1 - \varepsilon_0 L)\mathbf{P}/\varepsilon_0$.

Since this must vanish,

$$\mathbf{P} = -\varepsilon_0 \mathscr{E}_0/(1 - \varepsilon_0 L). \tag{2.15}$$

To translate this result into the change in conductivity produced by the void needs rather careful thought, and is conveniently approached in stages by way of the analogy between conductivity and dielectric permittivity, ε. First let us ask what is the mean permittivity of a volume of free space in which a dielectric ellipsoid is suspended, with a principal axis along $\mathscr{E}$ so that it is polarized parallel to $\mathscr{E}$, acquiring a dipole moment $\mathbf{p}$. To be definite let the ellipsoid be suspended in the parallel plate capacitor of fig. 10, so that the applied field is V/X. In the empty capacitor the displacement vector at the plates is $\mathbf{D} = \varepsilon_0 \mathscr{E}$, but the ellipsoid distorts the field and alters the mean value D_x of the normal component. There is a general theorem to the effect that introducing a dipole p_x into the capacitor when the plates are held at a fixed potential difference changes the charge on the left-hand plate from $\varepsilon_0 AV/X$ to $\varepsilon_0 AV/X + p_x/X$. And since D_x is equal to the surface charge density it follows immediately that D_x is changed by ΔD_x, where

$$\Delta D_x = p_x/AX = \bar{P}_x, \tag{2.16}$$

in which $\bar{P}_x$ is the average dipole moment per unit volume, AX being the volume within the capacitor. This result does not depend on the presence of the capacitor plates, and may be generalized in the form

$$\Delta D_i = \bar{P}_i. \tag{2.17}$$

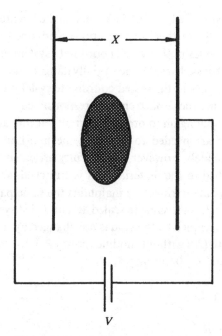

Figure 2.10 Calculation of the effect of a polarizable inclusion on the mean permittivity. The capacitor plates have area A.

Thus it is not necessary that $\bar{P}_i$ be parallel to the applied field $\mathscr{E}_j$, and we may write

$$\Delta D_i / D_j = \bar{P}_i / \varepsilon_0 \mathscr{E}_j \qquad (2.18)$$

or, since $D_i = \varepsilon_0 \varepsilon_{ij} \mathscr{E}_j$,

$$\Delta \varepsilon_{ij} / \varepsilon_{jj} = \bar{P}_i / \varepsilon_0 \mathscr{E}_j \quad \text{(no summing over repeated indices).} \qquad (2.19)$$

The analogy between dielectric and conducting materials is between $\mathbf{D}$ and $\mathbf{J}$, or ε_{ij} and σ_{ij}, so that we may immediately write the two special cases that concern us here:

1. If $\mathscr{E}$ and $\bar{\mathbf{P}}$ both lie along the z-direction (the case relevant, after some further considerations, to longitudinal magnetoresistance),

$$\Delta \sigma_{zz} / \sigma_{zz} = \bar{P}_z / \varepsilon_0 \mathscr{E}_z. \qquad (2.20)$$

2. If $\mathscr{E}$ lies along the x-direction, and $\bar{\mathbf{P}}$ in the plane normal to z (transverse magnetoresistance when σ_{xz} etc. $= 0$),

$$\left. \begin{array}{l} \Delta \sigma_{xx} / \sigma_{xx} = \bar{P}_x / \varepsilon_0 \mathscr{E}_x \\[2mm] \Delta \sigma_{yx} = -\Delta \sigma_{xy} / \sigma_{xx} = \bar{P}_y / \varepsilon_0 \mathscr{E}_x. \end{array} \right\} \qquad (2.21)$$

and

We can now return to the spherical void and its effect on the longitudinal magnetoresistance, and here the further considerations mentioned above come in, for the result (15) applies in the scaled coordinate system and we have to reverse the scaling process. There is a potentially dangerous hazard, in that Laplace's equation applies to the scaled coordinate system, so that there are no space charges, only the surface charges responsible for P. But when the pattern is stretched out again to produce current jets fore and aft of the cavity, space charges are needed to maintain the non-Laplacian potential distribution. Fortunately, however, no difficulty arises, since the equipotentials remain attached to the material as it is stretched, and the distribution of $\mathcal{E}_z$ over any plane normal to z maintains the same pattern. As a result, the disturbance to J_z is the same in scaled and unscaled systems; also the fraction, f, of space occupied by the void is not changed by scaling, so that (15) may be taken into (20) without modification, fP in the former becoming $\bar{P}_z$ in the latter, and $\mathcal{E}_0$ becoming $\mathcal{E}_z$.

Hence

$$\Delta\sigma_{zz}/\sigma_{zz} = -f/(1 - \varepsilon_0 L), \tag{2.22}$$

or

$$\Delta\rho_{zz}/\rho_{zz} = f/(1 - \varepsilon_0 L). \tag{2.23}$$

But remember that $\varepsilon_0 L$ is the depolarizing factor for the oblate spheroid in the scaled system, and $(1 - \varepsilon_0 L)$ may be very much less than the value $\frac{2}{3}$ for the original sphere. The void therefore may increase the resistivity much more than one would expect from its volume, and this is because current jetting produces long, virtually current-free, dead spaces on both sides.

The form of $\varepsilon_0 L$ for an oblate spheroid whose thickness is $1/R$ times its diameter,

$$\varepsilon_0 L = R^2(q - \tan^{-1} q)/q^3, \tag{2.24}$$

where $q = (R^2 - 1)^{1/2}$, tends as $R \to \infty$ (flat disc) to $1 - \pi/2R$, and in this limit when $\sigma_{zz} \gg \sigma_{xx}$, (23) becomes

$$\Delta\rho_{zz}/\rho_{zz} \sim 2fR/\pi \tag{2.25}$$

which is $(2/\pi)f\omega_c\tau$ for a free electron metal. When $\omega_c\tau = 200$ in potassium, $\frac{1}{2}\%$ of voids suffices to change ρ_{zz} so much that the bubbles can no longer be assumed to act independently – their current jets overlap and interfere with each other. Of course, we assume that Ohm's law applies locally everywhere, so that σ_{ij} is definable. It is all too likely that when the mean free path is long enough to allow $\omega_c\tau$ to reach such a high value, l is already comparable to, or even considerably larger than, the diameter of the bubbles. The theory then obviously needs complete overhauling, but this has never been done.

Turning now to the second case of special interest, we consider the effect of a spherical void on the transverse magnetoresistance. Here, in the scaled model, the void appears as an oblate spheroid with $\mathscr{E}$ and $\mathbf{J}$ in the equatorial plane, but the Hall effect does not allow us to assume $\mathbf{P}$ to be parallel to $\mathscr{E}$. On the other hand, if there is no longitudinal–transverse coupling we may take it to lie in the equatorial plane, and find its magnitude and direction by considering the boundary conditions at two points, rather than three as the most general form of σ_{ij} would demand. Let $\mathscr{E}$ be $\mathscr{E}_{x0}$, parallel to the x-direction, and choose the points where the x and y axes cut the spheroid. If the depolarizing factor is L', very small when R is large, the internal field components are $\mathscr{E}_{x0} - L'P_x$ and $- L'P_y$; by Gauss' theorem, exactly as before, the field just outside the x-pole has components.

$$\mathscr{E}_x^{(x)} = \mathscr{E}_{x0} + (1 - \varepsilon_0 L')P_x/\varepsilon_0, \quad \mathscr{E}_y^{(x)} = - L'P_y, \tag{2.26}$$

while just outside the y-pole

$$\mathscr{E}_x^{(y)} = \mathscr{E}_{x0} - L'P_x, \quad \mathscr{E}_y^{(y)} = (1 - \varepsilon_0 L')P_y/\varepsilon_0. \tag{2.27}$$

The normal component of $\mathbf{J}$ must vanish at both poles, so that

$$J_x^{(x)} = \sigma_1[\mathscr{E}_{x0} + (1 - \varepsilon_0 L')P_x/\varepsilon_0] - \sigma_2\varepsilon_0 L'P_y = 0$$

and

$$J_y^{(y)} = - \sigma_2(\mathscr{E}_{x0} - L'P_x) + \sigma_1(1 - \varepsilon_0 L')P_y/\varepsilon_0. \tag{2.28}$$

On solving for P_x and P_y we find

$$P_x/\varepsilon_0\mathscr{E}_{x0} = [\sigma_2^2\varepsilon_0 L' - \sigma_1^2(1 - \varepsilon_0 L')]/D$$

and

$$P_y/\varepsilon_0\mathscr{E}_{x0} = \sigma_1\sigma_2/D, \tag{2.29}$$

where $D = \sigma_1^2(1 - \varepsilon_0 L')^2 + \sigma_2^2\varepsilon_0^2 L'^2$. It is not difficult to show that all round the outside, on the equatorial plane, $\mathscr{E}$ meets the surface of the void at a constant angle, so that $\mathbf{J}$ can indeed be made to run parallel to the surface everywhere. It is more tedious to treat points off the equatorial plane, and here we rely on the general analysis to save the labour.

From (21) it follows that

$$\Delta\sigma_1/\sigma_1 = f[\sigma_2^2\varepsilon_0 L' - \sigma_1^2(1 - \varepsilon_0 L')]/D$$

and

$$\Delta\sigma_2/\sigma_1 = - f\sigma_1\sigma_2/D. \tag{2.30}$$

A general theorem[18] for ellipsoids is that the three principal values of $\varepsilon_0 L$ add to unity, and hence $\varepsilon_0 L' = \frac{1}{2}(1 - \varepsilon_0 L), \varepsilon_0 L$ being given by (24). If we apply (30) to a free electron metal for which, following (1.16), $\sigma_2/\sigma_1 = - \gamma$ and $R^2 = 1 + \gamma^2$,

$$\varepsilon_0 L' = [(1 + \gamma^2)\tan^{-1}\gamma - \gamma]/2\gamma^3. \tag{2.31}$$

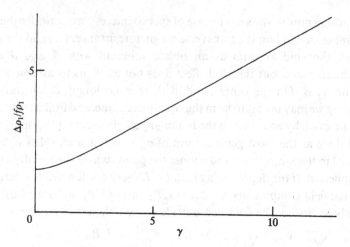

Figure 2.11 Effect of a fraction f of spherical voids on the transverse magnetoresistivity of a free-electron metal, according to (32).

To find the transverse magnetoresistance we note that $\rho_1 = \sigma_1/(\sigma_1^2 + \sigma_2^2)$, so that

$$\Delta\rho_1/\rho_1 = f[(\sigma_2^2 - \sigma_1^2)(\Delta\sigma_1/\sigma_1) - 2\sigma_1\sigma_2(\Delta\sigma_2/\sigma_1)]/(\sigma_1^2 + \sigma_2^2),$$

which simplifies for the free electron metal to

$$\Delta\rho_1/\rho_1 = f[(\gamma^2 - 1)\varepsilon_0 L' + 1]/[\gamma^2\varepsilon_0^2 L'^2 + (1 - \varepsilon_0 L')^2]. \qquad (2.32)$$

This is plotted in fig. 1.1. In zero field ($\gamma = 0$), $\Delta\rho_1/\rho_1 = 3f/2$, and as γ rises soon settles down to a linear increase:

$$\Delta\rho_1/f\rho_1 \sim 0.944 + \gamma/(4/\pi + \pi/4) = 0.944 + 0.486\gamma. \qquad (2.33)$$

The disturbing effect of voids is slightly less for transverse ($0.486f\gamma$) than for longitudinal ($0.637f\gamma$) magnetoresistance.[15] In both the origin of the massive disturbance when γ is large is current jetting, which is obvious in the longitudinal case but not quite so obvious in the transverse. For the latter, one should note that in the scaled model the current may appear to be only slightly deflected to pass on either side of the flat disc that represents the void. The bending of current lines, however, small as it is, extends away from the flat faces over a distance comparable to the largest radius, i.e. R times the half-thickness, since the field of the charges on the surface cannot be more closely confined. On reversing the scaling the current lines are bent as far away as R times the void radius,[19] measured along $\mathbf{B}$, and there must be shape charges extending this far to effect the necessary field distortion.

In conclusion, let us remember that these results have been derived for a free-electron metal, and their validity must be critically examined in any application. Nevertheless, the magnitude of the effect produced by voids is likely to be less than uncertainties in the theory in almost every case where an attempt is made to explain magnetoresistance quantitatively. The exceptions are metals like potassium (and possibly aluminium and indium) where the magnetoresistance is small, and these are precisely the metals for which the assumptions of the theory are not too far from reality.

Thermal magnetoresistance and the Righi–Leduc effect

The disturbing effect of magnetic field fluctuations and sample vibration has been mentioned as a possible limitation in potential measurement. One way of getting round this is the radical solution of measuring the thermal rather than electrical resistance – or instead of the Hall effect the Righi–Leduc effect, its thermal analogue. When the temperature is low enough for inelastic scattering of electrons by phonons to contribute little to the resistivity in comparison with elastic scattering by impurities etc., the Wiedemann–Franz law holds rather well.[20] If κ is the thermal conductivity due to the electrons, which normally dominate the conduction of heat,

$$\kappa/\sigma T = \tfrac{1}{3}(\pi k_B/e)^2 = 2.44 \times 10^{-8} \ \text{V}^2/\text{K}^2. \tag{2.34}$$

There is no reason to doubt that the parallelism between κ and σ holds for the individual components of κ_{ij} and σ_{ij}, and usually also in the presence of a magnetic field. Thus if one measures the transverse temperature difference produced when heat flows in the presence of **B**, the appropriate component of thermal resistivity, W_{yx}, should be related to ρ_{yx} by the suitably modified form of (34), $\rho_{yx}/W_{yx}T = 2.44 \times 10^{-8} \ \text{V}^2/\text{K}^2$. Even if the resistivity is not purely residual, and (34) is not strictly obeyed, it should prove little more difficult to interpret measurements of κ or W than of σ or ρ; but we need not discuss this.

There is, as one might expect, a price to be paid to eliminate in this way the noise due to vibration, for otherwise thermal measurements would have been more widely used. The price is the difficulty of finding thermometers that are not affected by the magnetic field. Gas thermometers are sensibly unaffected, but cumbersome and slow; resistance thermometers are fast but subject to magnetoresistance. Lipson,[21] in his study of the Righi–Leduc effect in copper, obviated the problem by connecting his carbon resistance thermometers to the sample by way of long rods of pure aluminium, which shows little magnetoresistance. He was then able to hold his thermometers well clear of the field without any apparent error in temperature measurement. The results will not be discussed here, but it is of interest to note the orders of magnitude, and indeed Lipson's paper repays study for its appraisal of experimental problems. First, the samples themselves could be made more bulky, and therefore much more robust, than for an electrical measurement. A matchstick of 5 mm square cross-section, carrying 1 A, gives a Hall voltage of about 10^{-7} V when $B = 5$ T; this was hard to measure to 1% at the time the work was done. By contrast, carbon thermometers had a sensitivity of 10^{-5} K and a difference of at least 10^{-3} K was desirable. According to (34), a heat flux as small as 1 mW should suffice, and in practice 10 times more could be passed without disturbing the cryostat as much as by 1 A through the sample.

One example of the value of thermal magnetoresistance, Fletcher's[7] study of the longitudinal effect in potassium, deserves close examination precisely because it reveals a huge discrepancy with (34). His method of attaching current electrodes has already been described, and the linear variation of ρ_{zz} shown in fig. 12 is typical of what several different groups have observed. There is no reason to doubt the efficacy of the bonding, and the explanation for the different behaviour of ρ_{zz} and W_{zz} must be sought in the sample material itself. If it were homogeneous, the Wiedemann–Franz law would demand that W as well as ρ rise linearly in the same proportion but, as Fletcher points out, the situation is different if there are voids. The scaling factor R that determines the value of $\varepsilon_0 L$ in (24) would be the same if σ_{zz}/σ_{xx} were equal to κ_{zz}/κ_{xx}. The lattice, however, contributes to the thermal conductivity, though only very weakly since the phonons that carry heat are heavily scattered by conduction electrons. In a pure sample of potassium, when $\mathbf{B} = 0$, the lattice contribution may be, at a very rough estimate, 10^{-4} of the electron contribution. Now $\sigma_{xx} \sim \sigma_0/(\omega_c\tau)^2$, and the electron contribution to κ_{xx} is similarly reduced by $(\omega_c\tau)^2$, so that $\omega_c\tau$ need only rise above 100 for the lattice contribution to dominate κ_{xx}. After this point κ_{zz}/κ_{xx} settles to a constant value and so also does W_{zz}. Fletcher finds fair, if not entirely satisfactory, quantitative agreement between his measurements and calculations based on this idea, but only by assuming the lattice thermal conductivity to be several times larger than accepted theory has frequently been thought to imply. The fault lies with naive interpretations of the theory, on the assumption that all phonons interact roughly equally with electrons. Strictly transverse phonons, however, are much less readily damped than any that possess a longitudinal component of vibration, and when the wave-vector lies in a plane of mirror symmetry there is always one strictly transverse phonon.[22] The reason is that, however anisotropic the crystal may be elastically (and potassium is very anisotropic) the three principal modes of vibration for a given wave-vector

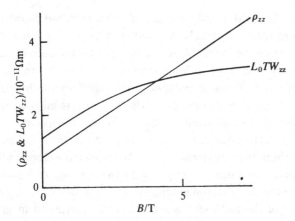

Figure 2.12 Longitudinal magnetoresistance, both electrical (ρ_{zz} at 4 K) and thermal (L_0TW_{zz} at 2 K) for a very pure potassium sample (Fletcher[7]). At the right $\omega_c\tau \sim 400$.

have mutually orthogonal displacements. In a mirror plane symmetry demands that one shall be normal to the plane, and therefore transverse; the other two lie in the plane and there is nothing to prevent both possessing longitudinal components of displacement. These latter contribute rather little, but the transverse mode, with its much longer free path, and the nearly transverse modes on either side of the mirror plane, are capable of dominating the conductivity.

This explanation of Fletcher's results is very persuasive but one must not assume too lightly that voids are the mechanism responsible for bringing in the current-jetting process. Any inhomogeneity that deflects the current lines from the z-direction will serve just as well[6] and lead to something like a linear rise of ρ_{zz}, accompanied by saturating W_{zz}. At this point we leave the potassium problem until chapter 5.

Electrodeless methods

It is possible to avoid the distortions of current flow near the electrodes in matchsticks by inducing currents in electrically isolated, and incidentally much more robust, samples such as spheres of single crystal material. Moreover the crystal axes can be disposed at a variety of directions relative to **B** without making a new sample. In principle a single spherical sample can cover all orientations, for which many matchsticks would be required;*

---

* Klauder *et al.*[2] covered a wide range of orientations with a single sample, by not insisting on strictly transverse **B**; this was entirely acceptable for their purpose (see Fig. 3.17).

in this way the difficulty of standardizing samples for mutual comparison, always a problem with soft metals, is avoided. Against this considerable advantage must be set the inevitable disadvantage that the current cannot take a unique direction as in a homogeneous matchstick sample. The interpretation of results is consequently not so straightforward. Indeed, the general theory of any one of these methods, when σ_{ij} is allowed its full complexity, shows that so intricate a mixture of components controls the measured quantity as to make one doubt its value. Nevertheless, if one can safely use less than the full form of σ_{ij}, by neglecting longitudinal–transverse coupling for instance, or going to the extreme of basing the interpretation on the free-electron model, much useful information may be deduced. Several of the methods discussed here are reviewed in greater detail by Delaney and Pippard.[4]

1. Eddy current decay

The simplest electrodeless method involves suddenly switching a small magnetic field b_z on or off, to induce eddy currents which then die away, in their turn inducing an EMF in a pick-up coil (which may be the same as the exciting coil).[23] If there is no longitudinal–transverse coupling the currents circulate in the x–y plane, with no z-component, and with a spherical sample their decay is governed by the mean resistivity in this plane, $\rho_t = \frac{1}{2}(\rho_{xx} + \rho_{yy})$. By applying a steady field B_z the mean transverse magnetoresistance can be measured.

There are many modes which decay exponentially, each with its characteristic time constant,[24]

$$\tau_n = \mu_0 a^2 / \pi^2 \rho_t n^2, \tag{2.35}$$

for a sphere of radius a. If b_z is uniform the magnetic moment of the sphere evolves according to

$$|M_z| = (12a^3 b_z / \pi \mu_0) \sum_{n=1}^{\infty} n^{-2} e^{-t/\tau_n}. \tag{2.36}$$

The lowest mode, $n = 1$, survives longest and analysis of the measurements is greatly eased if one waits long enough, after switching b_z on or off, for the higher modes to decay. In a sphere of radius 7.4 mm and with $\rho_1 = 4 \times 10^{-12}\,\Omega$m (aluminium with a residual resistance ratio $r_0 = 6000$), $\tau_1 = 1.7$ s, as in the tail of the curve in Fig. 13. In zero field the sample had $r_0 = 25\,000$, and the curve was taken with a strong field B present, giving $\Delta\rho/\rho = 3.2$.

It is obvious from the figure that (36) does not tell the whole story.

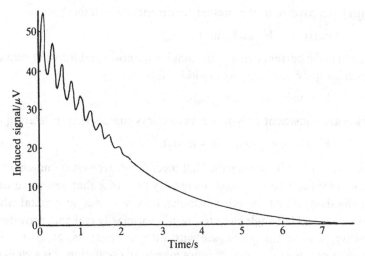

Figure 2.13 Induced EMF from an aluminium sphere (Grossbard[25]).

Oscillations like those at the beginning were first observed by Simpson[26] and shortly afterwards explained by Ford and Werner's[27] analysis of higher helicon modes in a sphere. Precisely which mode is revealed here is somewhat doubtful, but it is probably the lowest longitudinal helicon. This is irrelevant, being only a nuisance to be avoided by waiting long enough before measuring the decay. Helicons, however, provide another valuable electrodeless method and must be discussed in some detail.

2. Helicons

The physics of helicons involves many complications, few of which concern the process of measurment very closely. A major review, with very full bibliography, has been given by Petrashov.[28] Much of what we need is revealed by examining the propagation of electromagnetic waves in an infinite medium along the direction of **B**. The theory is virtually identical to that of magneto-ionic waves[29] in the ionosphere, but we are concerned here with a very much stronger field and a very much higher density of conduction electrons. As a result the velocity of propagation, instead of being of similar magnitude to the velocity of light, is reduced in a metal to only a few cm/s.

It is convenient to use complex numbers to describe the transverse field components and the conductivity. In writing Maxwell's equations the displacement current can be neglected as perhaps only 10^{-20} of the conduction current, and the field vectors in a plane wave, **e** and **b** (both entirely transverse in the present treatment) are related by

$$\text{curl}\,\mathbf{e} = -\,\mathbf{b} \quad \text{and} \quad \text{curl}\,\mathbf{b} = \mathrm{i}\mu_0\mathbf{j}, \tag{2.37}$$

where $\mathbf{j}$ is the current density. In complex notation and for wave motion in which all quantities vary as $\exp \mathrm{i}(kz - \omega t)$,

$$ke = \mathrm{i}\omega b \quad \text{and} \quad kb = \mu_0\rho e, \tag{2.38}$$

which are consistent only if the wave obeys the dispersion relation

$$k^2 = \mathrm{i}\omega\mu_0\sigma = \omega\mu_0\sigma_2(1 + \mathrm{i}\sigma_1\sigma_2). \tag{2.39}$$

In the absence of $\mathbf{B}$, or when the Hall effect is not strong, the imaginary term $\mathrm{i}\sigma_1/\sigma_2$ ensures a large enough imaginary part of k that any wave will be heavily damped. At the other extreme, however, in a pure metal when $\mathbf{B}$ gives rise to a Hall angle close to $\pi/2$, k^2 is primarily real and, provided it is positive, a wave can propagate with little attenuation. Note that when complex notation is used $\mathrm{e}^{-\mathrm{i}\omega t}$ is not merely an oscillation, it is a clockwise-rotating vector; and $\mathrm{e}^{\mathrm{i}\omega t}$ anticlockwise rotating. Depending on the sign of σ_2, that is on whether electrons or holes dominate the Hall effect, the propagating wave is either clockwise or anticlockwise circularly polarized. When it is reflected to form a standing wave the sign of k is reversed, but not that of ω. The standing wave therefore consists of half-wave sections between nodes in which the field strength at any one point stays constant in magnitude, but the field pattern as a whole spins about $\mathbf{B}$ as axis. If σ_2 is given by $1/\rho_{yx}$ in (1.59), i.e. by $e(n_+ - n_-)/B$, and we choose aluminium for an example, with $n_+ - n_- = 6 \times 10^{28}\,\mathrm{m}^{-3}$, the frequency is related to the wavelength, $\lambda = 2\pi/k$, by $\omega = 3.3 \times 10^{-3}\,B/\lambda^2\,\mathrm{s}^{-1}$ (B in Tesla, λ in metres). In a flat slab 1 cm thick, with $B = 5\,\mathrm{T}$ normal to the plane, the fundamental mode has $\lambda = 2\,\mathrm{cm}$ and its frequency is $41\,\mathrm{s}^{-1}$ or $6.5\,\mathrm{Hz}$.

The decrement of the free oscillation follows immediately from (39) for if $\omega = \omega' + \mathrm{i}\omega''$, $\omega''/\omega' = |\sigma_1/\sigma_2|$ which is $\pi/2$ minus the Hall angle when $\sigma_2 \gg \sigma_1$. For the sample of aluminium quoted above as having $\rho_1 = 4 \times 10^{-12}\,\Omega\mathrm{m}$, $\sigma_2/\sigma_1 \sim 130$ and the amplitude decays by a factor e in 21 cycles ($Q = \sigma_2/\sigma_1 = 65$). The higher modes of oscillation of the slab have successively 2, 3, etc. half-wave loops and their frequencies rise as the squares of the integers. The magnetic field vector, rotating about $\mathbf{B}$, causes each half-wave loop to behave as a rotating transverse magnetic dipole, with alternating signs in successive loops. The odd modes have a net moment which can be picked up by induction in coils wound round the sample (now assumed finite in lateral extent!), but the even modes are less strongly coupled, except to coils that respond with opposite sign to successive loops, e.g. a figure-of-eight for the second mode. It may be mentioned here, in anticipation of a later point, that the net moment of the odd modes interacts with **B** to generate a torque on the sample, whose axis spins with the helicon oscillation; the even modes do not so interact.

By treating standing waves resulting from the reflection of travelling waves in a laterally unbounded slab we have avoided the problem of boundary conditions at the sides, which is a much more awkward problem than in some other forms of standing wave. Among the first careful investigations of helicons in thin plates the experiments of Chambers and Jones[30] deserve a mention for their clear account of procedure and their analysis. They found the lowest resonance in several simple metals (Na, K, In, etc.) to occur at a frequency some 5% lower than expected, and it was only later that Legéndy[31] and others drew attention to the boundary value problem. Although Chambers and Jones had recognized, and attempted to allow for, the requirement that the transverse component of current must vanish at the sides, they had not taken account of the stored energy in the induction field outside the sample. Legéndy's calculations indicated that this might well account for the 5% discrepancy, but anything like a complete analysis had to wait for Klozenberg *et al.*[32] who, in a very intricate treatment, showed how complicated the behaviour of a circular cylindrical rod might be when **B** is along its axis. The fact that σ_{zz} is so much larger than the transverse components makes it easy for surface current jets to run parallel to **B** and generate a substantial magnetic field outside. There is every reason to be cautious about drawing conclusions in too great detail about the Hall conductivity from helicon frequency measurements. In addition the losses cannot be entirely attributed to transverse currents, and the simple relationship between σ_2/σ_1 and Q noted above is not to be trusted. The nearest one may hope to get to achieving this simplicity is to operate in a high mode of a flat slab, such that the wavelength is very much smaller than the lateral dimension and boundary effects are correspondingly less important. Chimenti and Maxfield[33] used the fiftieth harmonic, but still found a discrepancy of around 5%; this was in potassium, however, that notorious rogue among metals, and we must remember the observation when trying to make sense of its behaviour, rather than taking it as too convincing evidence against the reliability of helicons for measuring purposes.

In a spherical sample the surface current jets do not occur, since there is no extended surface, as on a cylinder, lying parallel to **B**. On the other hand there is still an induction field outside and a nasty boundary value problem, which was challenged by Ford and Werner[27] using brute force. They expanded the field patterns inside and outside the sphere in terms of 40 basis functions (vector spherical harmonics), naturally using a computer to solve the 40 × 40 matrix equations that resulted. In the course of this process they had to consider the symmetries involved and were led to tabulate the various transverse and longitudinal modes, their frequencies and decrements. In connection with the decrement it is worth noting that the Q-values of different modes may differ considerably, even for a given value of σ_2/σ_1, thus confirming that the different components of σ_{ij} enter into Q in a far from simple mixture. Thus for a free-electron metal, showing no magnetoresistance, the leading terms when $\omega_c\tau \gg 1$ are:

$$
\left.
\begin{aligned}
\text{in the lowest transverse mode,} \quad & \omega = 4.9020\,B/\mu_0 nea^2 \\
\text{and} \quad & Q = 0.4069\,\omega_c\tau(1-2.64/\omega_c^2\tau^2 + \cdots) \\
\text{in the lowest longitudinal mode,} \quad & \omega = 21.781\,B/\mu_0 nea^2 \\
\text{and} \quad & Q = 0.5058\,\omega_c\tau(1-81.5/\omega_c^2\tau^2 + \cdots)
\end{aligned}
\right\}
$$

(2.40)

The first coefficients in Q are $\sqrt{3}$ times those quoted by Werner and Ford, to allow for their unconventional definition of Q in terms of amplitude, rather than energy, decrement. Both are different from $\frac{1}{2}$, the value expected for pure transverse fields and currents.

Like the even modes in a flat slab, the longitudinal modes of a sphere have no transverse moment to couple with **B** and produce a torque. It is not so important to mount the sample rigidly to prevent movement, and for this reason Ford and Werner recommended using the lowest longitudinal mode to measure σ_2. This advice was followed by O'Shea and Springford[34] who found σ_2 in potassium to differ from the free electron value by less than others had reported, and indeed were not satisfied there was any discrepancy that could not be explained by sample imperfections, e.g. imperfect sphericity. Much the same conclusion is reached by de Podesta and Springford,[35] who find σ_2 in a field of 8 T to take the expected value within $(1.3 \pm 1)\%$.

3. Soft helicons

Before the longitudinal modes of a sphere had been discovered, when the lowest transverse mode was the only one considered to be available, Delaney and Pippard[36] proposed, to cope with the difficulty of mounting a sphere firmly enough without damaging it, to mount it on an elastic support and to make allowance for its movement in interpreting the results. The spinning magnetic moment, interacting with **B**, causes the sphere to rock in synchronism and to acquire extra induced currents from this movement in the field. This leads to a lowering of the natural frequency, and the resulting oscillation is called a *soft helicon* in contrast to the *hard helicon* in a firmly-held sample. This device probably invalidates the use of the helicon for measuring σ_2 absolutely, since the stiffness of the support must be known in order to correct for it, and the correction is liable to be large. On the other hand, we shall see immediately that measurements of σ_2/σ_1 are not disturbed, and the soft helicon has therefore some potential value in magnetoresistance studies.

The theory is simple in principle, especially if we assume that the stiffness of the support, C, is isotropic. Let us introduce a (complex) susceptance χ to describe the relation between a small field $be^{-i\omega t}$, rotating in a plane normal to $\mathbf{B}$, and the resulting induced magnetic moment $Me^{i\omega t}$ treating the transverse plane as a complex plane we write $M = \chi b$, in which $\chi = \chi(B, \omega)$ for a given sample. Now a rotating moment M, interacting with B, produces a torque MB which twists the sphere through an angle $\phi = |M|B/C = |\chi b|B/C$ in the plane containing $\mathbf{M}$ and $\mathbf{B}$. As far as the sphere is concerned, tilting it through ϕ in the direction of $\mathbf{B}$ has the same effect as tilting $\mathbf{B}$ through ϕ in the direction of $\mathbf{M}$. When ϕ is small this is equivalent to supplementing $\mathbf{B}$ by a transverse field $\mathbf{b}'$, parallel to $\mathbf{M}$, of magnitude $B\phi$. In complex notation, if b is real b' has the same orientation as M, and

$$b' = B^2 \chi b / C. \tag{2.41}$$

In free oscillation b' must itself play the role of b in generating M; consistency demands that

$$\chi = C/B^2, \tag{2.42}$$

and the frequency of oscillation is such as to achieve this.

When the sphere is clamped rigidly so that C is extremely large, χ must be correspondingly large, as it is at the natural helicon frequency. Otherwise a

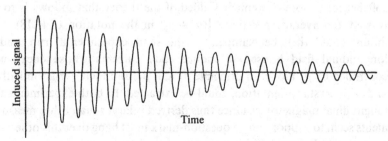

Figure 2.14 Chart recording of a soft helicon in the same sphere as used for fig. 13, at 4 K (Grossbard[25]). The oscillation period is 5 s.

smaller value of χ is called for, of the same sign, and this demands a lower frequency. Indeed if the mounting is quite flexible the natural frequency becomes very low (e.g. a period of 5 s for the aluminium sphere, in a field of 4 T, responsible for the trace shown in fig. 14). The advantage is that χ is much more easily calculated at such low frequencies that $\chi = A\omega$, with A a real constant when $\sigma_1 = 0$ and there are no resistive losses. There is the extra advantage that, since the right-hand side of (42) is real, the phase-angle of ω, i.e. ω''/ω', is immediately interpreted as the phase-angle, with sign reversed, of A. As (48) will show, in certain circumstances A is rather simply related to σ_{ij}, and its phase-angle gives information about σ_{xx} or ρ_{xx} that is not so reliably discovered by the use of hard helicons, since no theory has been given by Werner and Ford[27] except for a free electron metal.

Before proceeding to calculate A let us note some magnitudes. When ω is so high that the rotating field b is virtually precluded by the skin effect from entering the sphere, $|\chi| = 2\pi a^3/\mu_0$. Let us write this as χ_0 and take it as a characteristic measure of the magnitude of χ at other frequencies. If the helicon is to be only very little affected by the flexible mount, ω must be so near the resonance frequency that $|\chi| \gg \chi_0$. Otherwise expressed, (42) shows that this only occurs if C is much greater than the field energy defined as $B^2/2\mu_0$ times the volume of the sphere. For a field of 4 T and a sphere of radius 1 cm, as for the sphere used in fig. 14, $C \gg 27$ Nm/rad – this is the stiffness against bending of a steel rod of diameter 5 mm and length 24 cm. Clearly the use of large samples like this, that can be machined accurately and handled without damage, demands extremely (one might say prohibitively) robust equipment if the true hard helicon frequency is to be approached. For fig. 14, C was only 1.2 Nm/rad and the helicon was genuinely soft.

Systematic application of soft helicons by Simpson[26] (K) and Grossbard[37] (Al) was accompanied by the eddy current decay method.[23] With the sphere flexibly mounted, a transverse exciting field b initiated a soft helicon whose decrement yielded, if the theory that follows is to be trusted, the average resistivity $\frac{1}{2}(\rho_1 + \rho_3)$ in the notation of (1.18). The mount could then be clamped, in the same experimental run, and a longitudinal field b used to induce eddy currents whose decay was controlled by ρ_1 alone. In principle ρ_1 and ρ_3, and their variation with **B** and with crystal orientation, could be extracted. In both experiments the longitudinal magnetoresistance thus derived is larger than other measurements seem to support, and a question-mark must hang over this otherwise attractive technique.

The low-frequency susceptibility of a sphere

By low frequency we mean that the field **b**, which by rotating in a plane induces eddy currents, turns so slowly that the field of the eddy currents themselves is not enough to distort **b** significantly. One may then assume **b** to enter uniformly into the sphere and the pattern of currents to be independent of ω, having a magnitude and producing an induced moment proportional to ω, as assumed in the last section. It is often useful to imagine the direction of **b** to be fixed, along the z-axis, while the sphere rotates at angular velocity ω about an orthogonal axis which we take to be the x-axis. In the limiting case of a metal so pure that collisions can be neglected the electrons are prevented by **b** from following the rotation of the ionic lattice, which in any case is hardly entraining them by collisions. If there are n free electrons per unit volume, the charge density of the ionic lattice is $-ne$; a sphere carrying this charge density produces a moment parallel to the x-axis:

$$M_x = 4\pi \, ne \, \omega a^5/15. \tag{2.43}$$

The formula aplies equally well when **b** is not a weak field (in fact, when there are no collisions no field, however small, is weak since $\omega_c \tau$ is infinite), and we may generalize (43) by using the strong field limit (1.49) to give

$$M_x = 4\pi b_z \sigma_{xy} \omega a^5/15, \tag{2.44}$$

or

$$\chi_{xz} = 4\pi \sigma_{xy} \omega a^5/15 = 4\pi \, \omega a^5/15\rho_2. \tag{2.45}$$

In the limit where there are no losses **M** has no component in the plane normal to x, and there is no torque restraining the rotation of the sphere. Once one introduces losses, in small amounts at first so as not to disturb the current pattern very much, there must be a component M_y so that the work done in rotating the sphere, $\omega b_z M_y$ in unit time, matches the resistive dissipation. The moment tilts away from the x-axis to achieve this, and if the tilt is small that is the only significant alteration in the pattern. The angle of tilt is easily calculated from the loss-free current pattern,

$$J_0 = \omega ne(0, -z, y) \quad \text{or} \quad (\omega b_x/\rho_2)(0, -z, y). \tag{2.46}$$

The resulting rate of heat production is $\rho_1(J_x^2 + J_y^2) + \rho_3 J_z^2$ per unit volume, or $(\omega^2 b_x^2/\rho_2^2)(\rho_1 z^2 + \rho_3 y^2)$. Since the averages of x^2, y^2 and z^2 on a shell of radius r are all $\frac{1}{3}r^2$, we have, for the total rate of heat production,

$$\dot{Q} = \left[\frac{4\pi}{3} \omega^2 b_z^2(\rho_1 + \rho_3)/\rho_2^2 \right] \int_0^a r^4 dr = 4\pi \omega^2 b_z^2 a^5(\rho_1 + \rho_3)/15\rho_2^2. \tag{2.47}$$

Equating this to $\omega b_z M_y$ gives

$$M_y = 4\pi\omega b_z a^5 (\rho_1 + \rho_3)/15\rho_2^2, \quad \text{or}$$

$$\text{Im}[A] = 4\pi a^5 (\rho_1 + \rho_3)/15\rho_2^2, \tag{2.48}$$

and for the tilt of the moment away from the axis of rotation towards y,

$$\phi = M_y/M_x = (\rho_1 + \rho_3)/\rho_2, \tag{2.49}$$

which is twice the Hall angle for a free-electron metal.

This preliminary analysis serves to exhibit the physical process, so that the formal treatment[38] need not appear so repulsive. In general, when all components of σ_{ij} or σ_{ij} are present, the current pattern still maintains the form of (46), circular lines of flow, all in parallel planes, but there is no simple generalization for the orientation of these planes, analogous to (49). The component of moment M_y, which is responsible for the torque about the axis of rotation and is what is normally measured, always takes the form of (48):

$$M_y = (4\pi\omega b_z a^5/15) F(\rho_{ij}). \tag{2.50}$$

If longitudinal–transverse coupling may be neglected,

$$F = (\rho_{yy} + \rho_{zz})/[(\rho_{xx} + \rho_{zz})(\rho_{yy} + \rho_{zz}) - \rho_{xy}\rho_{yx}], \tag{2.51}$$

but in general,

$$F = [(\rho_{xx} + \rho_{yy})(\rho_{yy} + \rho_{zz}) - \rho_{xz}\rho_{zx}]/D,$$

where

$$D = (\rho_{xx} + \rho_{yy})(\rho_{yy} + \rho_{zz})(\rho_{zz} + \rho_{xx}) - \rho_{xy}\rho_{yx}(\rho_{xx} + \rho_{yy})$$
$$- \rho_{yz}\rho_{zy}(\rho_{yy} + \rho_{zz}) - \rho_{zx}\rho_{xz}(\rho_{zz} + \rho_{xx})$$
$$- \rho_{xy}\rho_{yz}\rho_{zx} - \rho_{xz}\rho_{zy}\rho_{yx}. \tag{2.52}$$

When $\omega_c\tau$ is so large that $\rho_{xx}, \rho_{yy} \ll \rho_{xy}$, (51) tends to the form (48) which is simple enough to be useful in principle, as the examples already quoted show. But one is sceptical about the possibility of making sense of observations that require (52) in its entirety for their interpretation. Visscher and Falicov[44], however, have shown how to extract useful information about the occurrence of magnetic breakdown in Zn and Cd.

Experimental applications of rotating field/sphere

In most applications of the foregoing theory the field b_z has been the total applied field B. The simplest form, pioneered (like most of the techniques mentioned in this section) by Datars and his school,[39] involved hanging the sample, originally a plate, later a sphere, on a torsion head equipped with feedback and torque recording; the horizontal field from an electro-magnet was slowly rotated by turning the magnet. For fields greater than 3 T, where a superconducting magnet is employed, it is usually easier to turn the sample and its associated measuring gear while keeping **B** fixed. There are variants according to the form of the magnet. With a superconducting pair in something like the Helmholtz configuration,[37] a horizontal magnetic field allows the sample to be turned continuously about a vertical axis, in a conventional helium cryostat. To obtain the strongest fields, however, a solenoid of small bore is needed and continuous rotation about an orthogonal axis is difficult, especially as extremely smooth rotation is needed if irregular torques are to be avoided. Verge, Altounian and Datars[40] have devised a compact design in which the sphere is rocked on a horizontal axis in a vertical solenoid, and the oscillatory torque measured with high sensitivity. Their account, and other references, give adequate constructional details for further description to be unnecessary here.

It is worthwhile, however, to comment on certain features of the rotation technique. For the first point it is necessary to anticipate the results of the next chapter, where it is shown that there are two mechanisms that lead to the transverse magnetoresistance, as measured on a matchstick sample, rising steadily as B^2 – compensation and open orbits. In a compensated   105 metal the high-field variation of ρ_{xx} and ρ_{yy} is as B^2, while ρ_{zz} tends to a constant value and $\rho_{xy} = -\rho_{yx} \propto B$, as shown by (1.60); ρ_{xz}, etc. may be ignored. Then (51) shows that as B rises F ultimately varies as B^{-2} and the torque on the sphere tends to a constant. When open orbits are present,   107

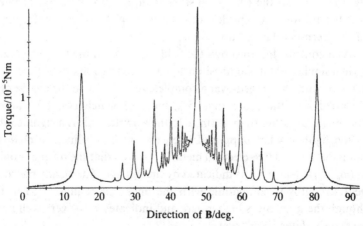

Figure 2.15 Torque on a tin crystal at 4 K as **B** turns uniformly in a [100] plane (Dixon[41]). The spikes indicate the presence of periodic open orbits.

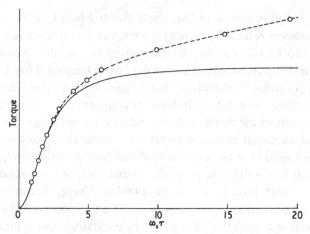

Figure 2.16 Torque on a potassium sphere rotating in a steady field **B**, at 4 K (Lass and Pippard[42]). The points represent the average torque over a complete revolution, and the full line is the theoretical curve for a free-electron metal.

there is a plane containing **B** in which conduction is relatively easy, and to which the circulating currents are drawn. Unless the normal to this plane lies close to the axis of rotation the resulting moment produces a strong torque, rising as B^2. The rotating sphere, unlike the matchstick, distinguishes between compensation and open orbits. Moreover, it is easy to give the sphere any orientation and so with a single crystal plot out all the directions of **B** relative to the crystal axes in which open orbits occur. An example of this mapping procedure applied to a tin crystal is shown in fig. 15; obviously there is a great store of information here, but we may leave it to the author and to professional Fermiologists to account for it in terms of the Fermi surface of tin.

As a contrast fig. 16 shows the field variation of the torque (at constant ω) in a single crystal of potassium, as measured by Lass.[42] At each value of **B** the torque is averaged over a complete revolution, in the course of which it varied by little more than 5%, most of which could be accounted instrumental rather than intrinsic to the sample. This is a significant result in the light of what happened later, as will be outlined in chapter 5. If potassium were a free-electron metal (51) shows that the torque should vary as $(\omega_c\tau)^2/[4 + (\omega_c\tau)^2]$, as indicated by the full line. The steady rise at higher field strengths is compatible with the linear effect illustrated in fig. 1.3, though the gradient is higher here and indicates a not very well prepared crystal (by later standards).

Let us turn now to the hazards besetting the use of this technique, many of which are discussed more fully by Delaney and Pippard[4] than can be reproduced here. Perhaps the most insidious arises if one forgets that when the Hall angle is near $\pi/2$ the component of **M** parallel to the axis of rotation is very much larger than the measured component. Any departure from perfect rigidity in the mounting is liable to cause the sample to twist, for instance by flexing the rod on which it is hung; a very small twist can entirely alter the measured torque. It is also possible that an oscillatory torque, due to de Haas–van Alphen oscillations of magnetization, can introduce apparent resistivity oscillations, all too easily interpreted as the Shubnikov–de Haas effect.

What was not available when this review was written, and deserves exposition, is the analysis of the disturbances due to non-sphericity of the sample. Lass[43] has shown that these can be serious, but his treatment is entirely mathematical and far from transparent physically. Let us therefore quote his result and then show in more physical terms how it comes about. One special case will suffice, a spheroid of free-electron metal, having axes $1, 1$ and b, the last being set normal to the rotation axis. In a given field, at constant ω, the moment M_y and the torque vary with the angle ϑ between **B** and the axis of length b according to:

$$M_y \propto (4 - 2\eta \sin^2 \vartheta + f\eta^2 \omega_c^2 \tau^2 \sin^2 2\vartheta)/(1 - h\eta \cos^2 \vartheta), \qquad (2.53)$$

in which

$$\eta = 1 - 1/b^2, \quad f = \tfrac{1}{2}b^2/(1 + b^2) \quad \text{and} \quad h = (2 + b^2)/(1 + b^2).$$

When $\omega_c \tau$ is not too large the angular variation is principally governed by $\eta \sin^2 \vartheta$ and $h\eta \cos^2 \vartheta$, giving rise to two maxima and minima per revolution. At large $\omega_c \tau$, however, $f\eta^2 \omega_c^2 \tau^2 \sin^2 2\vartheta$ may take control and then there are four maxima and minima per revolution. Since behaviour like this had been reported in some samples of potassium, it seemed probable that the explanation was to be found in imperfect sample shape. Later, however, samples that were unquestionably very nearly spherical showed the same pattern of maxima and minima (and indeed other more awful anomalies) so that the story was obviously incomplete. All the same, the result expressed in (53) is a clear warning to take great care of sample shape when $\omega_c \tau$ is large. Thus the last term in the numerator of (53) starts to dominate the first when $f^{1/2}\eta\omega_c \tau \sim 2$; if we write b as $1 + \varepsilon$, and $\omega_c \tau$ is large, this criterion can be written $\varepsilon \sim 2/\omega_c \tau$ and ε must be much less than this if it is not to introduce unacceptably large spurious variations of torque. Bearing in mind that in potassium $\omega_c \tau$ may easily exceed 200, we see that samples a few mm in diameter should be spherical to rather better than $10\,\mu m$.

To appreciate the origin of the terms in (53) we note first that **B** plays a part only in the last term of the numerator; the rest, which describes a slightly modulated moment, and therefore torque, with two cycles per revolution, can be understood in the absence of **B**, though the magnitude (contained in the proportionality constant) involves **B** through σ_{ij}. If the spheroid is prolate, with $b > 1, \eta$ is positive and the moment is largest when the b-axis lies along $\mathbf{B}(\vartheta = 0)$, smallest when it is normal to **B**. This is because as **B** turns relative to the spheroid, it is the xz-plane through which $\dot{\mathbf{B}}$ is greatest. Thus the greatest eddy currents are induced when the xz-section of the spheroid is maximum, in agreement with the form of these terms. Let us therefore take these as understood, in principle, and concentrate on the four-fold variation, proportional to $(\omega_c \tau)^2$.

It is obvious that if we ignore collisions and let $\omega_c \tau$ rise without limits, M_y also rises to infinity except when $\sin 2\vartheta = 0$. This is because in the absence of resistance **B** as it rotates must drag the electron assembly round with it as a rigid body – no problem in a sphere, impossible in a spheroid arranged to turn about an axis without axial symmetry. We must therefore keep collisions in our analysis, and ask how the electron assembly manages to conform to the external shape.

In fig. 17 a section of the spheroid is viewed along the axis of rotation of **B**; the axes Y and Z are fixed in the spheroid, while y and z rotate

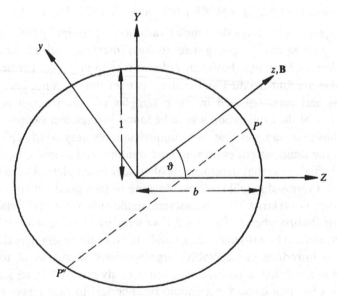

Figure 2.17 Notation for analysis of rotating spheroid.

anticlockwise with **B**. If we start by imagining the electron assembly to be rigid, the tangential current it produces lies in the plane and has density $ne\omega R$, where $R^2 = Z^2 + Y^2$; its outward-pointing component, normal to the boundary of the section, is $2ne\omega\varepsilon YZ/R$ and must be compensated by a non-divergent current in the spheroid which has an opposite normal component on the boundary. If ε is so small that terms in ε^2 may be neglected, these conditions are satisfied by the quadrupole current distribution.

$$\mathbf{J}_1' = -ne\omega\varepsilon(0, Z, Y).$$

Expressed in the rotating coordinates,

$$\mathbf{J}_1' = -ne\omega\varepsilon(0, z\cos 2\vartheta - y\sin 2\vartheta, z\sin 2\vartheta + y\cos 2\vartheta). \tag{2.54}$$

The electric field $\mathscr{E}_1'$ that $\mathbf{J}_1'$ generates is, however, not irrotational, as it must be since all induced fields have been dealt with by the rigid rotation of the electron assembly. At this stage only the leading terms in $\mathscr{E}_1'$ are needed, such as are generated by the Hall effect, and

$$\mathscr{E}_1' = -ne\omega\varepsilon\rho_2(z\cos 2\vartheta - y\sin 2\vartheta, 0, 0),$$

so that

$$\operatorname{curl}\mathscr{E}_1' = ne\omega\varepsilon\rho_2(0, \cos 2\vartheta, \sin 2\vartheta). \tag{2.55}$$

This takes the same value everywhere in the spheroid and we must supplement $\mathbf{J}_1'$ by $\mathbf{J}_1''$, non-divergent and not cutting the boundary, so as to eliminate $\operatorname{curl}\mathscr{E}_1'$. In a sphere the appropriate type of circulating current is shown in (46), in this case generating a moment M_x, and we generalize this to give an arbitrarily oriented moment, writing

$$\mathbf{j}_1'' = A_1(0, z, -y) + A_2(-z, 0, x) + A_3(y, -x, 0). \tag{2.56}$$

The coefficients A_1, A_2, and A_3 must be chosen to eliminate $\operatorname{curl}\mathscr{E}$. It may be objected that this current distribution, being conformable to a sphere, cuts the surface of the spheroid. The objection is valid if we hope to write $\mathbf{J}_1$ so precisely that we can evaluate M_y and hence the torque. We have already seen, however, how to avoid tedious precision by deducing the torque from the dissipation. Here, too, the approximation involved in neglecting the distinction between sphere and spheroid matters only in calculating quantities such as M_y, of first order in ε; but if ε is small enough to allow neglect of second-order terms, the dissipation yields the desired answer. By the same token, we must not expect to attach meaning to any but the leading term in our result, and may neglect anything that does not contribute to this.

When we come to evaluate A_1, A_2, and A_3 it becomes clear that ρ_1 as well as ρ_2 must be taken into account. It then turns out that the total current needed to provide the boundary fluxes is

$$\mathbf{J}_1 \approx - ne\omega\varepsilon \sin 2\vartheta [\tfrac{1}{2}\omega_c\tau(y, -x, 0) + (0, y, -z)], \tag{2.57}$$

in which ρ_2/ρ_1 is put equal to $\omega_c\tau$, and the only terms retained are those that contribute to curl $\mathscr{E}$ with a coefficient containing $\omega_c\tau$ (these, as arranged, cancel) and a small z-component so that div $\mathbf{J}_1 = 0$. So far as dissipation is concerned it is only the first vector field that matters; it represents a circulating current of order $\omega_c\tau$ greater than that in (54) which called it into being. Why this should be needed may be inferred from fig. 17; current injected at P' will jet, when $\omega_c\tau$ is large, along a line parallel to $\mathbf{B}$ and try to emerge at P'', where the boundary source in general will not match it. The current must therefore distribute itself in unpreferred directions. Only when $\vartheta = 0, \pi/2$, etc. does symmetry allow jetting to succeed without problems, and one can understand why the torque has four-fold rotational symmetry, with minima at these angles.

As for the magnitude of the dissipation, the circulating current in (57) has z as its axis and is orthogonal to the primary circulation about x. The two current sets operate independently in producing heat, and the extra component M_y that must be present follows by the same argument as led to (47). Comparing it with the primary moment M_y, we find

$$\Delta M_y/M_y = \tfrac{1}{4}\varepsilon^2\omega_c^2\tau^2 \sin^2 2\vartheta, \tag{2.58}$$

agreeing with the ratio of the first and last terms in the numerator of (53).

The dominant terms in (57) arise from the need to compensate for current sources due to sample imperfection, in this case asphericity. Such an undesired source may also come about through inhomogeneities, especially of ρ_2, which create current jets. Small-scale inhomogeneities have no effect other than those we have already discussed in the light of Herring's[13] analysis; large-scale inhomogeneities are hard to treat except in terms of artificial models, and this has not been attempted. It is clear that problems of this sort are no better dealt with in electrodeless methods than in matchstick samples. Indeed, one may say with confidence that no technique exists, or is likely to be invented, that will absolve the experimenter using high-$\omega_c\tau$ materials from subjecting his procedure and his results to the severest critical scrutiny; or, for that matter, persuading editors of journals to publish an adequate account of his procedure.

3 Real metals

Low-field magnetoresistance

The rise of semiconductor physics after 1945 stimulated interest in transport properties,[1] which offered the hope (largely realized) that they would lead to a correct description of the energy surfaces, followed by an understanding of how the scattering of electrons depended on their energy and their direction of motion. The Hall effect, which could be directly interpreted in terms of the carrier density, attracted most attention but other transport coefficients were not forgotten. In 1950 Seitz[2] showed that if σ_{ij} were expanded as a power series in **B**, with a view to characterizing the behaviour at low fields, the most general form allowed in a cubic crystal, up to the quadratic terms, could be written

$$\mathbf{J} = \sigma_0 \mathscr{E} + \alpha \mathscr{E} \wedge B + \beta B^2 \mathscr{E} + \gamma(\mathscr{E} \cdot \mathbf{B})\mathbf{B} + \delta T \mathscr{E}, \tag{3.1}$$

in which σ_0 and the coefficients α–δ are independent of **B** but, of course, temperature dependent*; T is a 3×3 matrix, diagonal when referred to the cube axes, with the form $T_{ij} = \delta_{ij} B_i B_j$ (no summation). The first three terms of (1) describe a conductivity tensor that is isotropic with respect to the orientation of **B**, the fourth is relevant to longitudinal but not to transverse magnetoresistance, while the fifth introduces anisotropy of the transverse effect. By a suitable choice of measurements all the coefficients may be determined; but how are they to be interpreted in terms of the electronic structure and scattering? There is no difficulty in expressing them in terms of a generalized model, though the result is hardly likely to inspire divination of the parameters of the model, which are many more in number

---

* It is assumed that the metal is non-magnetic. Iron and nickel are cubic but do not obey (1); in particular the Hall effect can be large and strongly non-linear if there is spontaneous magnetization. The theory of this *anomalous Hall effect* has been the subject of prolonged discussion, and has grown complicated and difficult in the process. Since it is only peripheral to our theme it will be ignored, with a recommendation to the potential enthusiast to start with Hurd's[3] survey and bibliography.

than the five independent quantities that experiment can reveal. Seitz chose a simplified model for germanium in which the energy surfaces were assumed spherical, while the relaxation time τ depended on energy and direction of motion: $\tau = \tau_0(E) + \tau_1(E)Y$, Y being a spherical harmonic with cubic symmetry. This gave him adequate flexibility to fit the data, without an embarrassing excess of freedom. Although at the time it was worth doing as a test of the procedure, the outcome is dubiously trustworthy. It is better to rely on other data, not to mention theoretical models, and use σ_{ij} as a firm rock on which to anchor the interpretation, rather than a springboard.

18   One valuable offshoot of this endeavour was Shockley's[4] tube integrals, the beginning of a geometrical approach to electronic structure. We have already met several examples and will find it most helpful, indeed virtually irreplaceable, when we come to discuss the high-field magnetoresistance of real metals. And the treatment of the low-field effects is made more physically intelligible by relating it to the shape of the Fermi surface rather than by starting with an arbitrary energy-band structure $E(\mathbf{k})$. To lead into this analysis let us examine one serious attempt to derive electronic structure from low-field measurements, Olson and Rodriguez'[5] attack on copper which, being cubic, must conform to (1).

A more convenient alternative to (1), when the measurements on matchstick samples relate to ρ_{ij}, is the general expression for $\mathscr{E}$ in terms of $\mathbf{J}$ and $\mathbf{B}$, up to second order in $\mathbf{B}$, which is quoted in a notation only slightly modified from that of Olson and Rodriguez,

$$\mathscr{E} = \rho_0 J [\mathbf{j} + (a'B/\rho_0)\mathbf{j} \wedge \mathbf{h} + (b'B^2/\rho_0^2)\mathbf{j}$$
$$+ (c'B^2/\rho_0^2)(\mathbf{j}\cdot\mathbf{h})\mathbf{h} + (d'B^2/\rho_0^2)\mathbf{T}'\mathbf{j}], \qquad (3.2)$$

where $\mathbf{j}$ and $\mathbf{h}$ are unit vectors defining the directions of $\mathbf{J}$ and $\mathbf{B}$, and $T'_{ij} = \delta_{ij}h_ih_j$. This way of writing the terms takes advantage of Kohler's rule which implies that in samples differing in the amount, not the kind, of scattering the coefficients $a'-d'$ should be sample-independent; and B/ρ_0 is a measure of $\omega_c\tau$. In copper, the assumption of a spherical Fermi surface (which is not too bad) leads through (1.3) to $B/\rho_0 = 1.38 \times 10^{10}\omega_c\tau T\Omega^{-1}\mathrm{m}^{-1}$. It is easy to relate (1) and (2) as follows:

$$a' = -\alpha' \quad , \quad b' = -(\alpha'^2 + \beta'^2) \quad , \quad c' = \alpha'^2 - \gamma' \quad \text{and} \quad d' = -\delta'$$
where
$$\alpha' = \rho_0^2\alpha \quad , \quad \beta' = \rho_0^3\beta, \quad \text{etc.}$$

Unlike Seitz's application to germanium, Olson and Rodriguez take τ in copper to be independent of $\mathbf{k}$ (for reasons which are plausible but unsound)

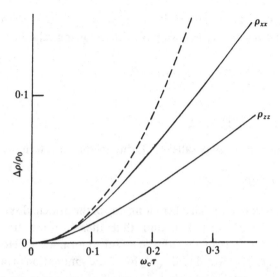

Figure 3.1 Magnetoresistance of polycrystalline copper at 78 K (de Launay *et al.*[1.4]), showing departure from quadratic behaviour (broken curve) at a low vlue of $\omega_c \tau$.

and parametrize the energy surfaces by the use of two spherical harmonics, and hence two adjustable parameters. They were fortunate in not having exceptionally pure copper at hand, so that they could survey the range in which $\omega_c \tau \ll 1$ without the embarrassment of running their electromagnet at very low currents. They found the quadratic range of magnetoresistance to fail noticeably when $\omega_c \tau$ was as low as 0.024 (a similar situation is shown in fig. 1, for a copper polycrystal at 78 K). With the benefit of hindsight they would have realized that this is evidence for rather sharp corners on the Fermi surface (cf. fig. 1.19) and offered this in support of their interpretation. For when they fitted the measured coefficients in (2) to their parametrized model, the resulting shape of the Fermi surface bulged enough to spread beyond the confines of the Brillouin zone; they suggested confidently that, in spite of theoretical predictions, the Fermi surface must make contact with the zone boundary at the centre of the (111) faces. It was their misfortune that a more direct determination of the Fermi surface,[6] with the same conclusion (see fig. 5), appeared only a few days later, and that their contribution has been undervalued. Later work[7] has confirmed the contacts beyond doubt.

Before leaving the formal aspects of (2) let us note its implications for anisotropy of the magnetoresistance and for longitudinal–transverse coupling. It is the last two terms that demand scrutiny, and if **J** and **B** are

37

orthogonal only the last one matters. Writing $\mathbf{j} = (j_1, j_2, j_3)$ with reference to the cube axes, and $\mathbf{h} = (h_1, h_2, h_3)$, we have the following relations:

$$\sum_{i=1}^{3} j_i^2 = 1 = \sum_{i=1}^{3} h_i^2 \quad , \quad \sum_{i=1}^{3} h_i j_i = 0, \tag{3.3}$$

and

$$\mathsf{T}'\mathbf{j} = (h_1^2 j_1, h_2^2 j_2, h_3^2 j_3),$$

so that the component of $\mathscr{E}$ parallel to $\mathbf{J}$, due to this fifth term, is

$$\mathscr{E}_{\|}^{(5)} = (d'JB^2/\rho_0) \sum_{i=1}^{3} h_i^2 j_i^2. \tag{3.4}$$

It is readily verified that $\mathscr{E}_{\|}^{(5)} = 0$ if $\mathbf{J}$ lies along any of the principal symmetry directions [100], [110] or [111], and that therefore the transverse magnetoresistance has the same quadratic low-field form for all. However, if we choose $[1, 1, \sqrt{2}]$ for $\mathbf{B}$ and $[\bar{1}, \bar{1}, \sqrt{2}]$ for $\mathbf{J}$, the summation in (4) takes its maximum value of 0.318, and unless d' is small, as it is not in copper, the term contributes significant anisotropy.

For the longitudinal magnetoresistance, $\mathbf{h} = \mathbf{j}$ and the fourth term in (2) is present but isotropic. Once more it is the fifth term that is responsible for anisotropy; its contribution to $\mathscr{E}$ may again be written as $\mathscr{E}^{(5)}$, with longitudinal and transverse components of magnitude

$$\mathscr{E}_{\|}^{(5)} = (d'JB^2/\rho_0) \sum_{i=1}^{3} j_i^4, \tag{3.5}$$

and

$$\mathscr{E}_{\perp}^{(5)} = (d'JB^2/\rho_0) \left\{ \sum_{i=1}^{3} j_i^6 - \left(\sum_{i=1}^{3} j_i^4 \right)^2 \right\}^{1/2} \tag{3.6}$$

$\mathscr{E}_{\|}^{(5)}$ is largest ($\sum = 1$) when $\mathbf{J}$ lies along a cube axis, and smallest ($\sum = \frac{1}{3}$) along [111]. In copper $d' > 0$ and the longitudinal magnetoresistance is least along [111], as is to be expected since the Fermi surface is most like a surface of revolution (which would show no magnetoresistance) about this axis.

5   The fifth term provides, through $\mathscr{E}_{\perp}^{(5)}$, the only source of longitudinal–transverse coupling, which must vanish by symmetry when $\mathbf{B}$ lies along a principal symmetry direction. Otherwise $\mathscr{E}_{\perp}^{(5)} \neq 0$ and takes its maximum value of 0.2806 when $\mathbf{B}$ and $\mathbf{J}$ lie close to $[1, 1, 3]$. At worst the longitudinal–transverse coupling is very nearly the same as the anisotropy of the magnetoresistivity. One must not be too ready to neglect it as a trivial nuisance.

Calculation of the low-field coefficients

The procedure developed in this section serves to illustrate how one may relate measured conductivity components to the shape of the Fermi surface and the variation over it of the mean free path, $l(\mathbf{k})$. We shall not proceed to the completion of the task since it will become clear that the expressions are too involved to allow an easy interpretation of σ_{ij} in terms of recognizable features of the electronic structure. Doubtless one could use the now-known shape of the Fermi surface in, say, copper together with the low-field components of σ_{ij} to attempt to find how $l(\mathbf{k})$ varies. But the answer would not be unique and no one, I think, has attempted such a programme.

The method is a variant of the impulse-response method used to derive (1.37). An impulsive electric field creates new particles on an element of the Fermi surface, and as they drift under the influence of $\mathbf{B}$, and disappear by catastrophic scattering, their contribution to the current density evolves and ultimately vanishes. The time-integral over this evolution determines the contribution of the surface element to the conductivity.

To define the surface element take a section δk_z thick, cut normal to $\mathbf{B}$, and a length δs_0 of the Fermi contour. If the normal to this element makes an angle θ_0 with $\mathbf{B}$, the area so defined is

$$\delta S_0 = \delta s_0 \delta k_z \operatorname{cosec} \theta_0. \qquad (3.7)$$

Let the impulsive field $\mathscr{E}_x$ act for time δt creating, according to (1.33), $N_0 \delta t$ particles where

$$N_0 = e\mathscr{E}_x \delta s_0 \delta k_z \cos \phi_0 / 4\pi^3 \hbar = e\mathscr{E}_x \delta S_0 \sin \theta_0 \cos \phi_0 / 4\pi^3 \hbar, \qquad (3.8)$$

ϕ_0 being the angle between $\mathscr{E}_x$ and the normal to the contour in the plane. The initial current density due to these new particles is

$$\delta^2 \mathbf{J}_0 = N_0 e \mathbf{v}_0 \delta t,$$

and at any subsequent instant, as $\mathbf{B}$ drives the bunch of particles round the contour and they decay in number by collision,

$$\delta^2 \mathbf{J} = N e v \delta t, \qquad (3.9)$$

We must now determine how the quantities in (9) evolve. First, from (1.25),

$$\dot{s} = \alpha v \sin \theta^*, \qquad (3.10)$$

while catastrophic scattering changes N according to

$$\dot{N} = -N/\tau = -Nv/l, \qquad (3.11)$$

---

* Do not confuse this $\alpha(=eB/\hbar)$ with the α in Seitz's notation of (1)!

in which τ and l are **k**-dependent, being the relaxation time and mean free path for an electron staying at a given **k** in the absence of **B**. Hence, from (10) and (11),

$$dN/ds = -(N/\alpha l)\operatorname{cosec}\theta. \tag{3.12}$$

Next we write the steady current generated by $\mathscr{E}_x$ acting on δS, by integrating (9) over t, and using (10) to write the result in the form

$$\delta \mathbf{J} = Ne \int_{s_0}^{\infty} (\mathbf{v}/\dot{s})\,ds = (\hbar/B)\int_{s_0}^{\infty} N(\mathbf{v}/v)\operatorname{cosec}\theta\,ds,$$

or, expressed in components,

$$\delta \mathbf{J} = (\hbar/B)\left[\int_{s_0}^{\infty} N\cos\phi\,ds, \int_{s_0}^{\infty} N\sin\phi\,ds, \int_{s_0}^{\infty} N\cot\theta\,ds\right]. \tag{3.13}$$

The reduction of (13) to useful form, as a power series in B, can be illustrated in general terms by evaluating $\int_{s_0}^{\infty} NF(s)\,ds$, $F(s)$ being any function of s. We convert N into dN/ds by means of (12), and integrate by parts:

$$\int_{s_0}^{\infty} NF\,ds = -\alpha \int_{s_0}^{\infty} Fl\sin\theta(dN/ds)\,ds = \alpha F_0 l_0 N_0 \sin\theta_0$$

$$+ B\int_{s_0}^{\infty} NF^{(1)}\,ds, \tag{3.14}$$

where $F^{(1)} = (e/\hbar)\,d\,(Fl\sin\theta)/ds$. In writing (14) we have made use of the vanishing of N as t, and therefore s, go to ∞. By defining a series of functions $F^{(n+1)} = (e/\hbar)\,d\,(F^{(n)}l\sin\theta)/ds$, we may continue the process to obtain the power series

$$\int_{s_0}^{\infty} NF\,ds = \alpha l_0 N_0 \sin\theta_0(F_0 + BF_0^{(1)} + B^2 F_0^{(2)} + \cdots), \tag{3.15}$$

the subscript zero indicating that each function is to be evaluated at the starting point, the surface element δS_0.

Let us apply this process to the coefficient of $\mathscr{E} \wedge \mathbf{B}$ in (1), describing the leading term in the Hall conductivity σ_{yx}, for which purpose (13) shows that we may choose $(\hbar/B)\sin\phi$ for F and evaluate δJ_y immediately:

$$\delta J_y = (\alpha^2 \mathscr{E}_x l_0 \sin^2\theta_0 \cos\phi_0/4\pi^3 B)\,dS_0$$

$$\times [\sin\phi_0 + \alpha d(l\sin\phi\sin\theta)/ds], \tag{3.16}$$

the derivative to be taken at s_0; in writing this N_0 has been replaced by use of (8). From this point we may drop the subscript zero. Since σ_{yx} is isotropic to first order in **B** we are free to choose cube edges as axes, and the Fermi contour has square symmetry. When we come to integrate over the Fermi surface to obtain the total J_y the only term in the derivative to survive is that whose symmetry is the same as $\cos \phi$; and we may replace $\cos^2 \phi$ by its average $\frac{1}{2}$. Then α (Seitz) is given by

$$\alpha(\text{Seitz}) = J_y/B\mathscr{E}_x = (e^3/8\pi^3\hbar^2) \int l^2 \, (d\phi/ds) \sin^3 \theta \, dS. \qquad (3.17)$$

This is still an awkward expression, involving a derivative on a contour which must be cut parallel to a cube face. But it can be manipulated into the form first given by Tsuji,[8]

$$\alpha(\text{Seitz}) = (e^3/24\pi^3\hbar^2) \int l^2 J \, dS, \qquad (3.18)$$

in which J is the *first curvature*[9] of the Fermi surface, i.e. the sum of the principal curvatures at each point (the same as determines the pressure difference across a curved surface due to surface tension). For a spherical surface $J = 2/k_F$, and with l constant (18) reduces to $\sigma_0 \omega_c \tau$ as required by (1.16). The manipulations that convert (17) into (18) follow in square brackets, and can be safely ignored.

[It is immaterial which cube face defines the section in which $d\phi/ds$ is to be measured, and we may replace $(d\phi/ds) \sin^3 \theta$ in (17) by the average over three orthogonal sections of the Fermi surface. Now $d\phi/ds$ is the local curvature in the section and if we take the point concerned as the origin of coordinates x_i, with the cube axes as coordinate frame, the surface can be expressed locally as a quadratic,

$$a_i x_i + b_{ij} x_i x_j = 0,$$

in which $a_i a_i = 1$, so that the a_i are direction cosines of the normal, and $b_{ji} = b_{ij}$. If we take a section through the origin, normal to x_1, $\sin^2 \theta = 1 - a_1^2 = a_2^2 + a_3^2$, and the equation of the curve is:

$$a_2 x_2 + a_3 x_3 + b_{22} x_2^2 + (b_{23} + b_{32}) x_2 x_3 + b_{33} x_3^2 = 0.$$

The curvature of the section is $(\partial^2 x_2/\partial x_3^2)/[1 + (\partial x_2/\partial x_3)^2]^{3/2}$, evaluated at the point concerned, i.e. the origin, and when this is done,

$$(d\phi/ds) \sin^3 \theta = 2[(b_{23} + b_{32})a_2 a_3 - b_{22}a_3^2 - b_{33}a_2^2].$$

On averaging over the three orthogonal sections, by cyclic permutation of subscripts, we find for the average value

$$\langle (d\phi/ds) \sin^3 \theta \rangle = \tfrac{2}{3} b_{ij}(\delta_{ij} - a_i a_j),$$

and this is $\frac{1}{3}$ times the first curvature J. Hence (18) follows.]

This derivation should serve to indicate the method to be followed, without going into detail, for the magnetoresistance effects which do not yield so neat an expression. Being quadratic in B they demand the evaluation of $F_0^{(2)}$ in (15), and, unlike the Hall effect, they are anisotropic. Thus if we choose the cube edges as axes to calculate the response J_x to a field $\mathscr{E}_x$ we arrive at the coefficient β in (1), and must leave the symmetry directions to find δ. Independent of any special choice of axes the calculation can proceed as before, but with an intermediate stage of integrating round the contour in a plane normal to $\mathbf{B}$. With the field $\mathscr{E}_x$ this leads to the quadratic term

$$\delta J_x = (e^4 \mathscr{E}_x B^2/4\pi^3\hbar^3)\delta k_z \oint l \sin\theta (dl_x/ds)^2 \, ds. \tag{3.19}$$

in which l_x is the x-component of $\mathbf{l}$, i.e. $l\sin\theta\cos\phi$. Another way of writing dl_x/ds is $\mathbf{h} \wedge \mathbf{V}_s l_x$, the operator $\mathbf{V}_s$ being a two-dimensional gradient lying in the Fermi surface. The first treatment of the problem, by Jones and Zener,[10] contained a somewhat fearsome operator Ω, defined as $(\mathbf{V}_k\varepsilon) \wedge \mathbf{V}_k$, in which $\mathbf{V}_k = (\partial/\partial k_x, \partial/\partial k_y, \partial/\partial k_z)$; since $\mathbf{V}_k\varepsilon = \hbar v$ and is normal to an energy surface, $\Omega = \hbar v\mathbf{V}_s$. None of this, however, makes the evaluation of (19) any easier. On substituting (7) and integrating over k_z, (19) becomes

$$J_x/\mathscr{E}_x B^2 = (e^4/4\pi^3\hbar^3) \int l \sin^2\theta(\mathbf{h} \wedge \mathbf{V}_s l_x)^2 \, dS. \tag{3.20}$$

Without a specific model in mind there is little point in trying to simplify this by applying the symmetry properties of the cubic system.

The corresponding expression relevant to longitudinal magnetoresistance is similar to (20),

$$J_z/(\mathscr{E}_z B^2) = (e^4/4\pi^3\hbar^3) \int l \sin^2\theta(\mathbf{h} \wedge \mathbf{V}_s l_z)^2 \, dS. \tag{3.21}$$

78   Again, it is hardly worth proceeding beyond this point. Olson and Rodriguez, as already remarked, set up a trial model with two adjustable parameters and carried the calculations to the point where they could be used. Since then, however, the low-field effects have attracted scant attention, and the best accounts are still those that belong to that period, such as the monographs of Wilson and Ziman.[11] These two books sit at the very boundary between two epochs in the history of the subject. For all the strength of his treatment, Wilson belongs to a somewhat formalistic school of thought that employed models of energy-band structure (e.g. two parabolic bands, one for electrons and the other for holes) that were susceptible to exact mathematical analysis; Ziman, by contrast, seeks to

approach the problems from the standpoint of realistic band structures, recognizing that in the end only computation will yield numbers to compare with experiment. It is the latter treatment that has taken over the field, as measurement and band-structure theory have united to give assured models to serve as a basis.

Survey of basic ideas

A general account of band-structure theory would be out of place here. Not only would it cover much more than is needed, but many treatments exist that are entirely adequate. It should perhaps be remarked that these frequently pass rather lightly over the fundamental difficulty of justifying the approach they take;[12] that is to say, they tend to take for granted that the electrons in a metal behave like independent particles, when in reality they interact among themselves with the strong Coulomb force, and indirectly by polarizing the ionic lattice (the process responsible for superconductivity). Nevertheless, subtle but convincing theoretical arguments and the success of experimental tests combine to establish the validity of the independent particle approach, subject to certain relatively small corrections which need not concern us. The ground state of the electron gas is not to be described by a product wave-function whose constituent members are the wave-functions of individual particles; on the contrary, it is a many-body wave-function of extreme complexity such as no-one would attempt to write down. When, however, one extra electron is added to the assembly, it must have energy at least as great as some characteristic value E_F, and when its energy is only slightly greater than this it behaves as an independent particle with a well-defined wave-number $\mathbf{k}$. The set of all $\mathbf{k}$ that are associated with an added electron, with energy E_F, defines the Fermi surface. It is in terms such as these that one should describe the situation that is summed up in the practical rules enunciated in chapter 1.

14

The constant-energy surfaces, especially the Fermi surface, that play the central role in magnetoresistance, and a short synopsis of their character will help to set the scene. It is frequently useful to think of them as periodic in $\mathbf{k}$-space, with the periodicity of the reciprocal lattice. To see the significance of the reciprocal lattice let us start with the real crystal, which may be generated by filling all space with replicas of a single unit cell, a parallelepiped defined by three basis vectors, $\mathbf{a}$, $\mathbf{b}$ and $\mathbf{c}$. If $\mathbf{R}$ is the vector joining any two equivalent points in the crystal, $\mathbf{R}$ is a linear combination of integral multiples of the basis vectors, $\mathbf{R} = l\mathbf{a} + m\mathbf{b} + n\mathbf{c}$. Now consider any function $f(\mathbf{r})$ which has the same periodicity as the crystal, so that $f(\mathbf{r} + l\mathbf{a} + m\mathbf{b} + n\mathbf{c}) = f(\mathbf{r})$, and express it as a Fourier series,

$$f(\mathbf{r}) = \sum A_i \exp(i\mathbf{g}_i \cdot \mathbf{r}). \tag{3.22}$$

The periodicity of $f(\mathbf{r})$ automatically restricts the vectors $\mathbf{g}_i$ that may appear in (22). For if we can add any $\mathbf{R}$ to $\mathbf{r}$ without changing $f(\mathbf{r})$,

$$\exp[i\mathbf{g}_i \cdot (l\mathbf{a} + m\mathbf{b} + n\mathbf{c})] = 1 \tag{3.23}$$

for all l, m and n. This is satisfied if $\mathbf{g}_i$ is a linear combination of integral multiples of $\mathbf{g}_a, \mathbf{g}_b$ and $\mathbf{g}_c$, which are defined so that

$$\mathbf{g}_a \cdot \mathbf{a} = 2\pi \quad , \quad \mathbf{g}_a \cdot \mathbf{b} = \mathbf{g}_a \cdot \mathbf{c} = 0, \tag{3.24}$$

and similarly for $\mathbf{g}_b$ and $\mathbf{g}_c$. Writing any permitted $\mathbf{g}_i$ as $\mathbf{G}$, we have that $\mathbf{G} = \lambda\mathbf{g}_a + \mu\mathbf{g}_b + \nu\mathbf{g}_c$, so that the set of all $\mathbf{G}$ form a periodic lattice, the reciprocal lattice, whose unit cell is formed from the basis vectors $\mathbf{g}_a, \mathbf{g}_b$ and $\mathbf{g}_c$. And only these vectors $\mathbf{G}$ can appear in the Fourier expansion of any function that has the periodicity of the crystal.

According to Bloch's theorem, the wave-function of an electron moving in the periodic potential of the crystal can always be written as

$$\psi_k(\mathbf{r}) = u_k(\mathbf{r})e^{i\mathbf{k}\cdot\mathbf{r}}, \tag{3.25}$$

in which $u_k(\mathbf{r})$ is not necessarily real, and has the periodicity of the crystal. Whatever form $u_k(\mathbf{r})$ may take, $\psi_k(\mathbf{r} + \mathbf{R}) = e^{i\mathbf{k}\cdot\mathbf{R}}\psi_k(\mathbf{r})$. Thus $\mathbf{k}$ defines the phase variation of ψ_k from one point to another that is its equivalent; it says nothing of local variations of phase and amplitude within the unit cell, which are the province of u_k. The discreteness of the information expressed by $\mathbf{k}$ has as a corollary a certain ambiguity in the definition of $\mathbf{k}$, since the same phase variation between equivalent points may be expressed by different $\mathbf{k}$. To see this, rewrite (25) in the form

$$\psi_k(\mathbf{r}) = u_k'(\mathbf{r})e^{i\mathbf{k}'\cdot\mathbf{r}},$$

in which $\mathbf{K}' = \mathbf{K} - \mathbf{G}$ and $u_k'(\mathbf{r}) = u_k(\mathbf{r})e^{i\mathbf{G}\cdot\mathbf{r}}$. Then $\mathbf{G}$ may be any reciprocal lattice vector without depriving $u_k'(\mathbf{r})$ of its essential periodicity in the real lattice. It follows that if an electron with wave-vector $\mathbf{k}$ has energy E, we are at liberty to associate an energy E with every point $\mathbf{k} + \mathbf{G}$. When we draw a surface of constant energy under these conditions it will be reproduced exactly in every cell of k-space as defined by the basis vectors $\mathbf{g}_a, \mathbf{g}_b$ and $\mathbf{g}_c$.

There is nothing sacred about the form of the unit cell in k-space (or in r-space for that matter). It may be mathematically economical to take it as a parallelepiped, but the symmetry of the lattice may thereby be obscured. Strictly, the only necessary condition on the unit cell is that it shall be capable of filling all space by periodic replication, and that any one point in

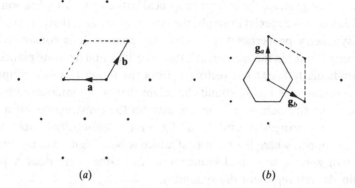

(a)   (b)

Figure 3.2 Hexagonal two-dimensional lattice (a) and its reciprocal lattice (b), with basis vectors and unit cells (completed by broken lines). The hexagonal Brillouin zone is also shown in (b).

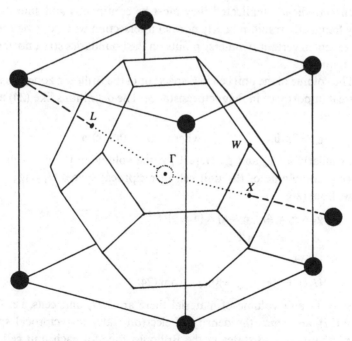

Figure 3.3 bcc reciprocal lattice and Brillouin zone for copper, with certain symmetry points labelled; X and L are midpoints of lines from the centre Γ to neighbouring reciprocal lattice points.

the cell shall generate the correct reciprocal lattice by such replication. The *Brillouin zone* is a special example, the most important in that it reproduces any symmetry properties the crystal may possess. It is constructed by choosing a point in the reciprocal lattice as origin, and drawing planes that perpendicularly bisect the vectors **R** from the origin to other reciprocal lattice points. The region round the origin that is not intersected by any plane is the Brillouin zone. Fig. 2 illustrates the construction for a two-dimensional hexagonal crystal, and fig. 3 for a face-centred cubic metal, such as copper, where the reciprocal lattice is body-centred cubic and the Brillouin zone a truncated octahedron. In neither case does a paral-lelepipedal cell represent the symmetry.

The faces of the Brillouin zones contain those values of **k** at which Bragg reflection of X-rays occurs, or of any other wave that interacts only weakly with the crystal. From the theory of weak interactions, which is frequently invalid for electrons in metals, there has sprung up a common misconception that energy surfaces must meet the Brillouin zone boundaries normal to them. In fact the necessary conditions to be satisfied by energy surfaces are less restrictive; they must conform to the symmetry of the crystal, and when periodically replicated they must be continuous and smooth. This may frequently result in nearly normal intersection with zone boundaries, but except at certain symmetry points on the boundaries strict normality is not required.

The volume of the unit cell in k-space, or of the Brillouin zone, is a matter of great importance in magnetoresistance. The definitions like (24) may be rewritten

$$\mathbf{g}_a = 2\pi(\mathbf{b} \wedge \mathbf{c})/U_r \quad , \quad \text{where} \quad U_r = (\mathbf{b} \wedge \mathbf{c}) \cdot \mathbf{a}, \tag{3.26}$$

and similarly for $\mathbf{g}_b$ and $\mathbf{g}_c$. Here U_r is the volume of the unit cell in real space. The volume of the unit cell in reciprocal space, $U_k = (\mathbf{g}_b \wedge \mathbf{g}_c) \cdot \mathbf{g}_a$. Now, from (26),

$$\mathbf{g}_b \wedge \mathbf{g}_c = 4\pi^2(\mathbf{c} \wedge \mathbf{a}) \wedge (\mathbf{a} \wedge \mathbf{b})/U_r^2$$
$$= 4\pi^2 \mathbf{a}/U_r.$$

Hence

$$U_k U_r = 4\pi^2 \mathbf{a} \cdot \mathbf{g}_a = 8\pi^3, \quad \text{from (24).} \tag{3.27}$$

Since in unit volume of material there are $1/U_r$ unit cells, i.e. $U_k/8\pi^3$ from (27), and since the density of electron states in reciprocal space is $1/4\pi^3$, there are two states in the Brillouin zone for each unit cell in the specimen of material. This means that a metal in which the unit cell contains one monovalent atom (Na, Cu) has enough conduction electrons to be

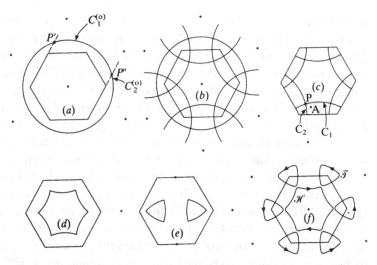

Figure 3.4 Harrison construction for a two-dimensional hexagonal metal, as described in text.

described by half the states available in a single Brillouin zone, while a divalent atom needs as many states as a Brillouin zone contains, or even twice as many if, like the hexagonal metals Mg, Zn, etc., there are two atoms per unit cell*. The face-centred cubic Al is trivalent, with one atom per unit cell, and has electrons to fill one-and-a-half zones. To accommodate these electrons several energy bands must be wholly or partially occupied; that is to say, for each $\mathbf{k}$ (and its replicas $\mathbf{k} + \mathbf{G}$) there is a set of wave-functions $\psi_{\mathbf{k}}^{(i)}$, each with its characteristic energy $E^{(i)}(\mathbf{k})$; and each band index (i) defines its own set of energy surfaces.

This is readily appreciated by applying the idea of periodic replication to a free electron metal, one that possesses a crystal lattice with which electron interactions are negligibly weak. The two-dimensional hexagonal structure of fig. 2 will serve our purpose. The reciprocal lattice and Brillouin zone are repeated in fig. 4(a), together with a circular constant-energy contour. To represent the information in the periodically replicated scheme we draw the same contour translated by vectors $\mathbf{G}$ – in this case, drawing the circle with every reciprocal lattice point as centre, as in (b). This is known as the Harrison[13] construction. There is no need to keep anything that lies

---

* It is the spin quantum number that is responsible for there being two states; if spin were absent the density of states would be $1/8\pi^3$. In non-magnetic metals the two spin states are equivalent, but one must be careful when ferromagnetic metals are being considered to remember that the Brillouin zone contains only one state of a given spin for each unit cell of the crystal.[14]

outside the chosen unit cell, and if this is the Brillouin zone we can be content with (c) as the basis for replication. Note, however, that the point A lies outside the circle that generates the arc C_2, and presumably represents a state of higher energy than that of the contour; but it lies inside the circle that generates the arc C_1, and by the same token represents a state of lower energy. This inconsistency is resolved by recognizing that the arcs are generated from portions $C_1^{(0)}$ and $C_2^{(0)}$ of the original circle by translating through different G, and that their intersection P describes two different wave-functions, plane waves from P' and P'' of the free-electron representation. The diagram only makes sense if we draw each branch in its own unit cell, and there is a formal procedure for carrying out the reconstruction systematically: to assign a *zone index* to any point on the contour, draw a circle (in three dimensions a sphere) centred on that point and passing through the origin, i.e. the centre of the free-electron sphere. The number of reciprocal lattice points so enclosed, counting the origin as one, is the zone index. Thus $C_1^{(0)}$ belongs to the second zone in which the energy contour is the curvilinear hexagon (d), but $C_2^{(0)}$ belongs to the third zone (e). This has been drawn with a different choice of unit cell, to illustrate the flexibility available and to show the continuity of the contours in the periodically replicated scheme. If there is any electron–lattice interaction the degeneracy at the point where arcs cut in (b) is removed, and the cusps are rounded off, as in (f), to give smoothly continuous contours. Strictly speaking, to remove the degeneracy of two plane-wave free-electron wave-functions, the lattice potential must contain a Fourier coefficient that matches the difference in wave-vector of the two states. At the intersection of two arcs in the Harrison construction the required Fourier coefficient always belongs to the set G, but it may be of small amplitude, in which case the splitting of the previously degenerate levels is small; or it may vanish entirely for symmetry reasons, and the degeneracy remain. The two zones then make contact at this point.

The contours in (f) are labelled with arrows to show how the magnetic field B would drive free electrons round the arcs – in this case, anticlockwise. The slightly rounded cusps are the points where Bragg reflection disturbs the motion fundamentally, causing electrons in the hexagonal orbit to execute a clockwise motion, as if they were positively charged. This reversal of direction always accompanies an inward-pointing velocity vector. If the contour corresponds to the Fermi energy, all states inside the original free-electron circle are filled and all outside are empty. In the reconstruction this results in all states between the hexagon and the zone boundary being filled, and all inside the hexagon empty. This is a *hole orbit*.

By contrast, the electrons run anticlockwise round the triangles, which contain filled states and are *electron orbits*.

If the lattice potential should be unable to remove degeneracy at a crossover, it will equally be powerless to cause Bragg reflection, and the electron will continue in its original circle. The transition from Bragg reflection to free-electron-like behaviour is the phenomenon of *magnetic breakdown*, whose occurrence is governed by the strength of the relevant component of lattice potential and by the magnitude of **B**. This will be the principal topic of chapter 4.

153

The small closed surfaces that are frequently generated by the Harrison construction, in Al, Mg, Zn for example, have correspondingly small cyclotron masses. It is easy to show, by use of (1.29), that the corners (provided **B** is not strong enough to cause magnetic breakdown) make little contribution to the time needed to complete an orbit. The cyclotron mass, therefore, is smaller than the mass in the unbroken orbit to something like the same extent that the perimeter is smaller. The very small mass already noted in bismuth may likewise be roughly attributed to the very small Fermi surface left after the lattice potential has broken up some hypothetical quasi-free orbit.

Typical Fermi surfaces

The examples that follow are chosen from those metals whose magnetoresistive behaviour will concern us most closely. Much more detailed accounts may be found in the book by Cracknell and Wong,[15] while Shoenberg[16] gives an up-to-date survey, with particular emphasis on the evidence obtained by means of the de Haas–van Alphen effect which is by far the most powerful tool for determining Fermi surface shapes.

Potassium: the Fermi surface is very nearly spherical, with variations of radius vector amounting to no more than 0.14% (for a conflicting view, see chapter 5).

191

Copper (and silver and gold which are similar): these are fcc metals, whose Brillouin zone is shown in fig. 3. The Fermi surface of copper is shown in fig. 5; it is drawn out along [111] directions to contact the hexagonal faces. Consequently, when replicas are stacked so that space is filled by the Brillouin zones, the Fermi surface has the appearance of a body-centred cubic array of spheres, joined by necks at their closest approach. Fig. 6(*a*) is a photograph of a crude model which, when sectioned by a (110) plane

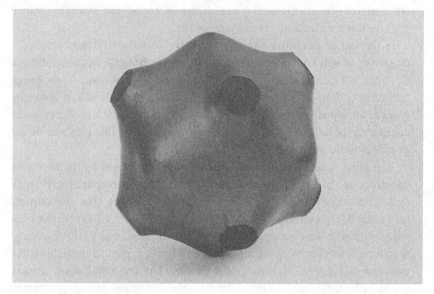

Figure 3.5 Model of the Fermi surface of copper, with orientation similar to fig. 3. The black discs, centred on points L, show areas of contact with the zone boundary.

through the centre (the plane through Γ, X and L in fig. 3), gives the hole orbits (*dogsbone* orbits) of (*b*). The true section (*c*) shows that the model is a fair representation. When **B** lies along [110] the electrons in this section describe dogsbone orbits in real space, oriented as usual at $\pi/2$ to the k-orbits. As we shall see, multiply-connected Fermi surfaces like this are capable of generating a remarkable variety of orbits as the orientation of **B**

99

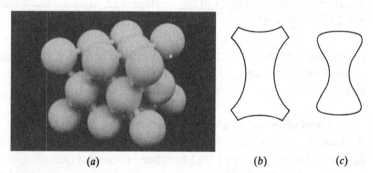

        (*a*)   (*b*)   (*c*)

Figure 3.6 Schematic model of the surface in fig. 5 after replication. The necks connecting neighbouring replicas appear sectioned, as black discs, in fig. 5. The forward-facing surface is a (110) plane, as is that facing upwards, and that facing to the right is (100).

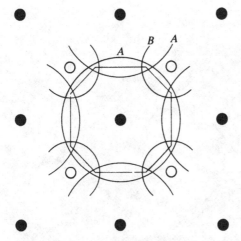

Figure 3.7 Section of the Fermi surface of aluminium, in the Harrison construction, after replication; the overlaps at the corners are exaggerated.

is changed, and the magnetoresistance is to a large extent controlled by their form.

Aluminium: this is also an fcc metal, but with three valence electrons per atom instead of copper's one. The Harrison construction starts with a sphere that runs close to the corners of the Brillouin zone, its radius being greater by only 0.8% than the distance from centre to zone corner. The (100) central section, through Γ, X and W in fig. 3, is shown in fig. 7. The larger circular arcs are sections of the spheres centred on reciprocal lattice points in the plane (shown as solid circles), while the smaller arcs are sections of spheres centred on the open points, which lie out of the plane of the section. The intersections, near the corners, of the two sets indicate that Fourier components of lattice potential corresponding to reciprocal lattice vectors [200] and [111] will both contribute to rearranging the corrections. Ashcroft's[17] examination of the different possibilities, depending on magnitude and sign of these components, has led him to a solution that fits the experimental data well. It may be remarked that this analysis is an excellent illustration of the power of the pseudo-potential[18] approach to band structure when, as in aluminium, the perturbation by the lattice potential is unexpectedly weak.

The major part of the Fermi surface occupies, as a hole surface, the zone of index 2, that of index 1 being filled by states of energy less than E_F. The Harrison construction for this part is shown by the model in fig. 8, on which the section drawn in fig. 7 is indicated. The sharp edges are only slightly

Figure 3.8 Model of the second-zone hole surface in aluminium; the heavy line indicates the section shown as the innermost orbit in fig. 7 (Souvenir of the Cooperstown conference[26]).

rounded by the lattice interaction. In the zone of index 3 is an arrangement of electron surfaces, each of which is a square torus like 4 sausages connected in a ring (fig. 9(*b*)); these roughly coincide with the square faces of the Brillouin zone, as illustrated by fig. 9(*a*).

To relate the electron surfaces to the free-electron Harrison construction, a single corner W is shown enlarged in fig. 10. The arrows from W to W' and W'' run along two sides of a square zone face, while the arrow from W to X runs diagonally across the other square that touches the first at W and lies at a right angle to it. These lines are shown in fig. 9(*a*). The broken lines in fig. 10(*a*) are sections of the free-electron spheres, lettered in conformity with fig. 7; B is therefore the cusped intersection of two spheres, whose degeneracy is broken to give B_2, a section of the second zone hole surface, and B_3, a section of the third zone electron surface. The triangular overlap at W is displaced, and A and B_2 reconnected, to give S, the section of the third zone surface at the corner of a square.

As drawn in fig. 10(*a*), the degeneracy at C is broken, but Ashcroft finds that the Fourier component of lattice potential fails to separate the second and third zones everywhere, so that they have a contact point as in (*b*); an electron lying in the particular section that contains this contact will suffer

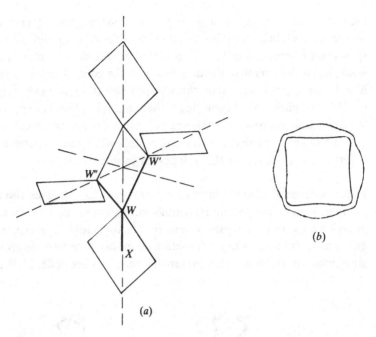

Figure 3.9 (*a*) Connections of the square faces of the Brillouin zone in fig. 3, after replication. In aluminium the third-zone Fermi surface has the form (*b*) and sits just inside a square face; in the replicated pattern each square is the site of such a ring. The heavier lines in (*a*) run from *W* to *W'*, *W"* and *X*; they serve to relate fig. 10 to this space pattern.

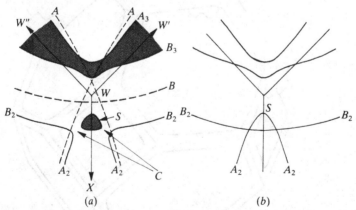

Figure 3.10 Detail of fig. 9 (Ashcroft[17]). The shaded parts of (*a*) are different sections of the corner of the third-zone electron surface, fig. 9(*b*). The two lines A_2B_2 are boundaries of two second-zone hole surfaces, as in fig. 8; in the absence of spin–orbit splitting they have contact points with *S*, as shown in (*b*).

magnetic breakdown, passing from one zone to the other without hindrance. As it happens this is not quite true, for Ashcroft did not include spin–orbit interaction in his treatment. In a light element like Al this is weak, but it does remove the degeneracy at the contact point so that (*a*) is a better representation of the electron orbits in a weak magnetic field.[19] In higher fields the splitting of the bands is insufficient to prevent an electron passing across, and (*b*) is nearer the truth; better expressed, there is a non-vanishing probability for orbits following either pattern. The experimental evidence for this will appear in due course.

174

Lead: this metal, also fcc, deserves a mention as having given rise to the point of view expressed in Harrison's construction, as well as having provided impetus for the pseudo-potential theory. Gold[20] interpreted his measurements in terms of the free-electron model, with four electrons per atom, which leads to the extended tubular Fermi surface of fig. 11. We shall

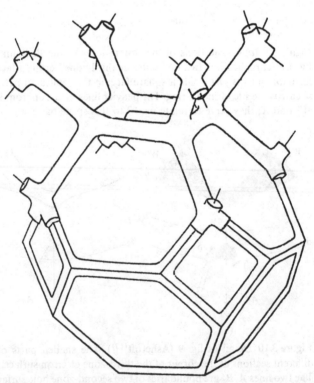

Figure 3.11 Schematic drawing of the third-zone electron surface in lead (Gold[20]).

not discuss the magnetoresistance of lead, but it may be noted that although the Fermi surface is multiply connected, as is that of copper, there is not here the variety of orbits that copper can generate; all sections of the tubular structure give electron orbits. There are also closed hole surfaces of equal volume.

Magnesium (and Be, Zn and Cd):   these are hexagonal metals, not far from being close-packed but with slightly different c/a ratios which influence their Fermi surfaces to a rather minor extent; the strength of the relevant components of lattice potential is more important. It is enough here to describe the Fermi surface of magnesium. The crystal structure consists of layers of atoms packed closely in a triangular pattern, but with alternate layers fitting into the spaces at the centres of the triangles. Strictly there are two atoms in the unit cell, but their environments are so similar that it is often good enough to draw a Brillouin zone in the form of a hexagonal prism twice as high as the correct zone and holding two electrons, rather than one, per atom. Since there are two valence electrons per atom, the free-electron sphere has the same volume as this zone, but of course it overlaps the boundaries to give distinct branches of the Fermi surface in zones of different band index. Much of the detail is of little concern to us

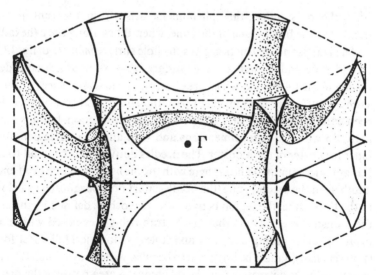

Figure 3.12 Harrison construction for the second-zone hole surface (monster) in magnesium. The Brillouin zone is shown by broken lines and its central plane by full lines.

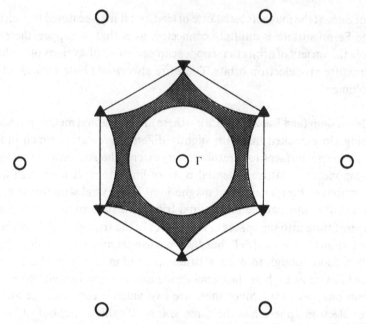

Figure 3.13 Central section of the monster of fig. 12, with reciprocal lattice points as open circles. The dark triangles are the third-zone electron surfaces (needles).

since the most interesting phenomena arise from the central section, parallel to the basal plane of the zone, when **B** lies along the c (hexad) axis.

The Harrison construction gives the hole surface *monster* of fig. 12 in the second zone and, as well as other surfaces that we need not consider, the *needles* in the third zone, which are electron surfaces and twice as numerous as the monsters in the replicated pattern. The central section has already been used in fig. 4 to illustrate the Harrison construction, but it is repeated in fig. 13 with the right dimensions and with an extra circle denoting the inner perimeter of the monster. The needles derive from the small triangles. It is not uncommon to find, as here with the monster, that one orbit encloses another. In the second zone (the needles must now be ignored as belonging to the third zone) the region between the two orbits, defined by the arms of the monster, is empty, so that an electron orbit is enclosed within a hole orbit. When counting electrons and holes, as for the Hall effect formula (1.59), each orbit must be imagined to be filled with its own kind; thus in this case $n_+ - n_-$ is determined by the difference in area between the orbits. In three dimensions, it is the volume within the monster that determines the excess of holes in this zone.

Varieties of orbit

The rule for finding how an electron will move in a magnetic field has already been derived, and can be applied to realistic Fermi surfaces without change, provided we employ periodic replication to complete any k-orbit which may not be contained within a single Brillouin zone. Sections of the Fermi surface by planes normal to **B** define the k-orbits, which are reproduced in real space after scaling by $1/\alpha$ and turning through $\pi/2$. This is to ignore motion along **B**, but for transverse magnetoresistance, which we shall concentrate on, this component of the motion is unimportant. The different types of orbit produced by sectioning are all illustrated by the Fermi surface of copper as represented in fig. 6, and we shall take this to demonstrate the principles.

If we were to choose a direction for **B**, represented by $[lmn]$, at random, almost certainly l, m and n would be mutually incommensurable, with no rational factor in common. In that case a plane normal to **B** that passed through one reciprocal lattice point would pass through no other, however far the replication was extended; and the plane would generate every possible section of the Fermi surface with this orientation. On the other hand, when the orientation of **B** is commensurable, especially if l, m and n are small integers or zero, a given plane only generates a limited number of sections. One must take care, in enumerating all possible orbits, not to overlook any when **B** lies along directions of high symmetry.

Let us start with such a direction, $[100]$, for which two types of orbit are found in copper, according to the level at which the section is made. If the plane cuts the midpoints of four necks a hole orbit results, as in fig. 14(a), and neighbouring planes give similar orbits with narrower neck regions. As

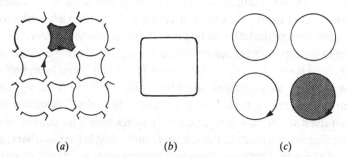

(a)   (b)   (c)

Figure 3.14 Sections of the Fermi surface of copper by (100) planes, showing (a) a hole orbit (shaded) as given by the model of fig. 6, (b) the true shape of this orbit and (c) an electron orbit (shaded) as given by the model.

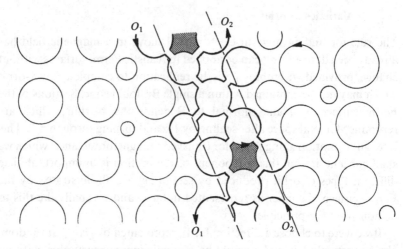

Figure 3.15 A belt of hole orbits (shaded) bounded by open orbits O_1 and O_2, when **B** lies near [100]. The arrows show the opposite directions of electron motion on the two open orbits.

(b) makes clear, these orbits are considerably more like a square than the model suggests. Other planes cut those parts of the Fermi surface which are uninterrupted by necks, and generate electron orbits, (c). The volume of the Fermi surface that gives electron orbits defines n_-, while the unoccupied volume in the slice that contains hole orbits defines n_+. There is no simple formula for n_-, n_+ or $n_- - n_+$; they must be computed from the shape of the surface.

When **B** is tilted away from [100], by however little, the situation changes radically, as was first made clear by Lifshitz and Peschanskii.[21] Fig. 15 shows schematically a section of the model in fig. 6, the plane being chosen to cut through the centre of a zone at the bottom left, and also at the top right, though the latter section is one layer above the former. In between these two electron-orbit regions there is a region where the plane cuts the necks, and is sufficiently little tilted from [100] that it produces hole orbits. Continued in all directions the plane would be traversed by strips of electron orbits interspersed with strips of hole orbits. If it were a map on which filled states were land and empty states water, we should recognize archipelagoes of islands in the sea, and landlocked lakes elsewhere; and we should know that between these two regions there must be a coastline. In the diagram two coastlines, O_1 and O_2, are marked more heavily, and are seen to consist of partial circuits of the Fermi surface that sometimes hit a neck and sometimes miss*. However sinuous one of these lines may be locally, its general direction follows a straight line determined by the intersection of a (100) plane of the crystal with the plane normal to **B**. For along this line all orbits have the same character. If **B** is kept normal to the paper, the pattern of fig. 15 is obtained by starting with a cube axis along **B** and tilting it about this line.

An electron whose state is defined by a point on a boundary line is impelled by **B** to travel along it indefinitely, if it is not scattered, and in real space to travel along a sinuous path that lies in the vicinity of a straight line normal to this. It may also have a mean component of velocity v_z, so that its general direction is straight but not necessarily normal to **B**. This is an *open orbit*, and it may be either periodic or aperiodic; the former if **B** is in a commensurate orientation, the latter if the orientation is incommensurate. Open orbits occur in pairs, like those in fig. 1.5, and the electrons move along them, propelled by **B**, in opposite directions.

As **B** is tilted away from [100] the belts of electron orbits and those of hole orbits are initially very wide, so that the fraction of electrons in open orbits starts at zero and increases as the greater tilt reduces the width of the belts. There comes a point, however, when the hole orbit belt becomes too narrow to sustain an open orbit; instead we find greatly elongated but finite closed orbits (*extended orbits*). This is illustrated in fig. 16 which simplifies the representation to its uttermost. The checkered pattern represents the projection of the bcc arrangement of filled states onto a (100) plane, the shaded areas being filled states inside the Fermi surface. Wherever a plane normal to **B** cuts a neck (shown as a point) the shaded areas are joined at their common corner, and where the neck is missed by the plane the corners are separated. If **B** drives the electrons anticlockwise round the edges of the shaded areas, they turn to the left at a corner where the neck is uncut, to the right where it is cut. For any orientation of **B** we may draw the edges of the belts within which the necks are cut, and proceed immediately to construct the orbits. The situation in (*a*) is similar to that in fig. 15 except that the tilt of **B** is increased and the hole orbit belt narrowed. The orbit *A* is a typical extended orbit, electron-like in character as are all extended orbits in copper, and *A'* is part of another. They are flanked by simple electron orbits.

---

* In Mg and Pb we find electron and hole orbits coexisting on the same plane, without the extended coastline of Cu, because one encloses the other like a lake on an island, or vice versa. Also, of course, when more than one zone is occupied there may be only electron orbits in one band and only hole orbits in another, but no coastline boundary between them since they occupy different planes.

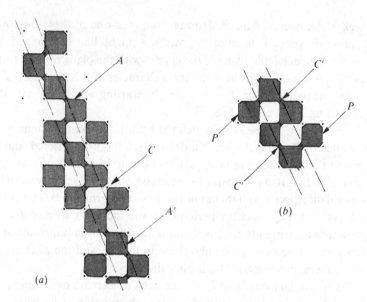

Figure 3.16(a) Schematic diagram of an extended orbit (*A*) in copper. In (*b*) **B** has been tilted slightly into a direction of low rational indices, and the orbit has turned into a periodic open orbit, of which only the repeat unit is shown.

If the shapes were correctly drawn according to the form of the real Fermi surface, and **B** were pointing in an incommensurate direction, the areas of different types of orbit on a very large expanse of the plane section would correctly enumerate the relative numbers of electrons associated with each. When, however, open orbits are present such enumeration is not possible, since an open orbit has the character neither of an electron nor of a hole.

In fig. 16(*b*) the direction of **B** has been chosen to be partially commensurate, i.e. [2*m*, *m*, *n*]. The tilt is roughly the same as in (*a*) but because the orbit pattern is now periodic, with the repeat pattern as shown, it is possible to find a section that lacks the closure seen at *C* in (*a*). If either of the necks indicated by *C'* were to lie outside the belt the orbit would be closed, with a transition to a row of quite short orbits. On the other hand, by making the direction of **B** incommensurate, for which an infinitesimal turning of the belt in the plane of the paper suffices, we cause the location of the belt, as we follow along it, to drift slowly across the checkerboard and encounter closure patterns like *C*. Thus if we start with **B** along [*l*, *m*, *n*] and with *l* = 2*m*, increase of *n* leads to the sudden replacement of periodic open orbits by short closed orbits, while changes to *l* or *m* alone can replace them

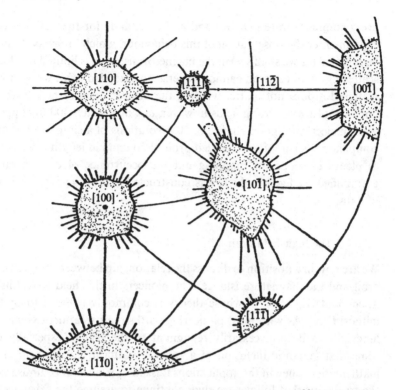

Figure 3.17 Stereographic representation (Klauder *et al.*[22]) of directions of **B** in which non-saturation reveals periodic open orbits in copper, as indicated by the lines, and zones (shaded) of mainly aperiodic open orbits. At [11$\bar{2}$] two sets of periodic open orbits coexist.

by long extended orbits. But this is only true when n is large enough, i.e. **B** far enough from [100]. Closer in there are always open orbits, periodic for commensurate and aperiodic for incommensurate directions.

This discussion should help to explain the remarkable complexity of the stereographic plot in fig. 17. At each of the principal symmetry directions, [100], [110] and [111], only small electron and hole orbits are to be found, as indicated by fig. 14 for [100] and fig. 6(*b*) for [110]. These are surrounded by quite wide solid angles in which there are open orbits, and from these regions jut out lines in commensurate directions where there are periodic open orbits flanked by highly extended orbits. The larger the integral indices needed to define the commensurate directions, the shorter their projection beyond the shaded open orbit regions. Indeed, as one approaches the shaded regions the spikes become ever more numerous until, at the moment of reaching the shade, they are infinitely close, representing every commensurate pair of l and m. Fortunately for the patience of the experimenter the observation of this behaviour, which in principle should be revealed (as we shall see) in the magnetoresistance, is limited by electron scattering. If an electron cannot traverse an extended orbit without being scattered it does not matter whether the orbit is open or not. With very pure samples and strong fields however, $\omega_c\tau$ may reach 300 in copper, to give an electron a good chance of 50 revolutions of a simple orbit, or the complete traversal of an extended orbit of 50 units in length. This provides plenty of scope for mapping out the occurrence of open orbits, as exemplified by fig. 17 which was constructed from a vast accumulation of data.

The high-field limit

We are now in a position to discuss the relationship between the varieties of orbit and the magnetoresistance. The pioneers in this field were Lifshitz, Azbel' and Kaganov[23] whose theory is commonly referred to by their initials, LAK. As with most pioneering works their treatment seems with hindsight to be unnecessarily repugnant to the reader, especially their somewhat pedantic inclusion of a general scattering function in all the mathematics, when in the applications it is either irrelevant or unusable. In the treatment that follows we shall continue to assume the existence of a relaxation time $\tau(\mathbf{k})$, pointing out later that this assumption excludes certain possible forms of behaviour that are more readily understood by qualitative physical reasoning than by the use of general scattering functions. This criticism of the presentation in no way detracts from their achievement in delineating the principal magnetoresistance effects before much data existed, and when little was known of the shape of real Fermi surfaces. A good survey of this aspect of the field, made at the time when it was an active research area, is to be found in a review by Fawcett.[14]

Much progress can be made by qualitative arguments, with the use of Kohler's rule and Onsager's relations. The former, as in (1.42), allows all components of σ_{ij}/σ_0 to be expressed as functions of a conventionally defined γ, the same for all electrons, irrespective of variations of ω_c and τ among them; γ is simply an arbitrary constant times B/ρ_0. The latter demands that a high-field expansion of σ_{ij} as a series in $1/\gamma$ shall contain only even powers in the diagonal terms, and that for the off-diagonal terms $\sigma_{ij}(\mathbf{B}) = \sigma_{ji}(-\mathbf{B})$.

1. Closed orbits

If all orbits are closed, there can be no constant term except in σ_{zz}, for in the absence of collisions ($\gamma = \infty$) no steady motion of electrons can take place except along z. Keeping only the leading terms we may write for the high-field behaviour ($\gamma \gg 1$)

$$\sigma_{ij}/\sigma_0 \approx \begin{bmatrix} a/\gamma^2 & d/\gamma & e/\gamma \\ -d/\gamma & b/\gamma^2 & f/\gamma \\ -e/\gamma & -f/\gamma & c \end{bmatrix} \qquad (3.28)$$

in which a–f are constants whose value does not concern us immediately. Hence, from (1.12),

$$\rho_{xx}/\rho_0 \approx b(1 + f^2/bc)/d^2, \qquad (3.29)$$

where once more only leading terms are retained. Provided the Hall effect does not vanish, as in compensated metals, $d \neq 0$ and ρ_{xx} saturates at high field strengths. This is what happens in figs. 1.4 and 1.5(*a*), if we are willing to overlook (as we shall for the moment) the slow linear rise after 'saturation' is complete – something that cannot emerge from an expansion which must contain for ρ_{xx} only even powers of B. Various explanations of how the linear rise can come about will be introduced at appropriate points, but it remains an unsatisfactory aspect of the subject.

It follows, by putting $c = 1$ in (1.54), that $d = 0$ for a compensated metal, but σ_{xy} may contain a leading term in $1/\gamma^2$ except when symmetry precludes this. Whether this is present or not, inversion of (28) always gives ρ_{xx} ultimately increasing as γ^2 i.e. as B^2. Compensation is not an unusual phenomenon since, according to (27), the Brillouin zone contains two electrons for each unit cell of the crystal. The conduction electrons in even-valenced metals can therefore just fill one or more zones. If they do so without spilling over into higher zones and leaving vacant states in lower zones, they are not metals at all but insulators at 0 K, and possibly semiconductors if thermal excitation produces a few conduction electrons (Ge, Si, etc.). Metallic conduction demands spilling over into higher bands, as already described in the case of magnesium, and then the number in higher zones matches the vacant states left behind. Quite commonly the latter constitute a closed Fermi surface of holes and the former a closed surface of electrons; then $n_+ = n_-$ and compensation occurs, with ρ_{xx} rising quadratically at high fields. This is the explanation for the large magnetoresistance in Mg and Zn, which will be discussed at length in chapter 4 since at higher fields magnetic breakdown interrupts the quadratic rise. The

30   quadratic effect in bismuth is also the result of compensation, as noted in chapter 1. In this case there are 2 atoms, and 10 valence electrons, in the unit cell so that compensation is possible in spite of the odd valency.

Before leaving closed orbits let us glance at some quantitative aspects of the magnetoresistance of copper when $\mathbf{B}$ lies along a high-symmetry direction. We have already noted that for [100], [110] and [111] there are closed electron and hole orbits; in addition, the coefficients e and f in (28) vanish by symmetry if the direction of current flow is also one of high symmetry, e.g. [011]. Then the saturation value of ρ_{xx}/ρ_0, according to (29), is b/d^2. We might assume b to be not greatly different from its free-electron value of unity; d, on the other hand, which would also be unity for a free-electron metal, must be replaced, according to (1.54), by $(n_- - n_+)/n_{\mathrm{f}}$, where n_{f} is the actual electron density. We therefore estimate the saturation value of ρ_{xx}/ρ_0 to be $n_{\mathrm{f}}^2/(n_- - n_+)^2$, and this is very readily calculated from the Fermi surface.

Consider the surface in fig. 5 and for convenience let the height of the Brillouin zone between square faces be 2 units, so that the volume of the zone is 4 and of the Fermi surface itself 2. When $\mathbf{B}$ lies along [001] the belt in the upper half defined by the necks, within which the holes lie, runs between $\frac{1}{2} \pm \zeta$, where ζ must be determined from the shape and is close to 0.105. Let us write V_1 for the volume within the Fermi surface defined by this belt, and

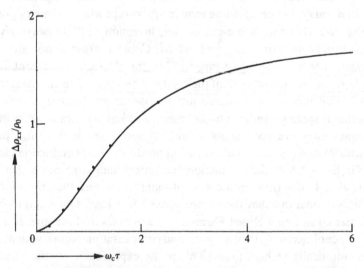

Figure 3.18 Computation of simplified model for copper with $\mathbf{B}$ along [100], comparing (30) (full line) with points taken from the measurements of Lengeler and Papastaikoudis[24].

V_2 for the volume of the zone similarly defined. Then the volume occupied by holes in the upper half is $V_2 - V_1$, and the volume occupied by electrons is $1 - V_1$. The difference, $1 - V_2$, compared to the actual volume of the Fermi surface, which is unity, gives $(n_- - n_+)/n_f = 0.58$. By this reckoning $\rho_{xx}(\infty)/\rho_0$ should be 2.97, or $\Delta\rho_{xx}/\rho_0$ not far from 2.

Surprisingly, in view of the extensive study that copper has undergone, no curve appears to have been published that allows us an immediate check of this estimate. There is, however, a curve[24] with **B** in this orientation which takes $\omega_c\tau$ as far as 3 and shows ρ_{xx} to be tending towards the expected saturation value. Let us accept the challenge of finding the shape of this curve with the simplest assumptions, viz. that τ is independent of **k**, that ω_c is the same for holes as for electrons (since the hole orbit in fig. 14 is made up of portions of a complete circuit of the surface), that the electron orbits are circular and the hole orbits square. We now use (1.16) to describe the electron conductivity and (1.86) and (1.87) for the hole conductivity, suitably normalized to give the required high-field Hall conductivity, proportional to $n_+ - n_-$. This leads to

$$\left.\begin{aligned}\sigma_{xx} = \sigma_{yy} &\propto (4n_+/\pi)[1 - (2\gamma/\pi)\tanh(\pi/2\gamma)] + n_-/(1 + \gamma^2) \\ \text{and} \\ \sigma_{xy} &\propto (8\gamma n_+/\pi^2)[1 - \operatorname{sech}(\pi/2\gamma)] - \gamma n_-/(1 + \gamma^2).\end{aligned}\right\} \quad (3.30)$$

From these expressions ρ_{xx} is easily computed as $\sigma_{xx}/(\sigma_{xx}^2 + \sigma_{xy}^2)$, with the result shown in fig. 18. There is one adjustable constant – the fraction of electrons that are unaffected by the necks and move in electron orbits. This has been taken to be 0.76 to get a good fit to the experimental points; and hence, as already noted, the fraction appearing as holes is $0.76 - 0.58$, i.e. 0.18. Measurement of the Fermi surface[25] indicates that 0.745 electrons and 0.165 holes are nearer the truth, but in view of the extremely simplified assumptions one can be satisfied that the theory, at least in this intermediate range of $\omega_c\tau$, is in good shape. This should not be taken as a highly critical test – the proportion of electrons and holes is the most important single factor, orbit shapes and variations of τ being relatively minor influences, as we saw in chapter 1.

2. *Extended and open orbits*

Extended orbits like those in fig. 16 are not different in principle from closed orbits, and the resistivity may be expected to saturate in the same circumstances as with small closed orbits. Since, however, they are made up of many elementary units, the cyclotron frequency is low and a correspondingly large field will be needed to reach saturation. Until this happens the resistivity continues to rise, so that $\Delta\rho_{xx}(\infty)/\rho_0$ may reach large values. And if the extended orbit turns into an open orbit saturation never occurs. It is easiest to start the discussion with open orbits, remembering that until the point of saturation is reached there is little difference between open and extended orbits.

Except when the Fermi surface has many separate sheets which generate open orbits or, in a simple metal like copper, for a small number of special high-symmetry directions, one finds only one set of open orbits at a time, oriented as described above by the intersection of a lattice plane with the plane normal to **B**. This determines the direction in which an electron moves under the influence of **B**, and when it collides only rarely it can travel through many units of the orbit. In fact, along the orbit direction its effective path **L** is comparable to the mean free path when **B** = 0, only reduced by a geometrical factor because of the sinuosity of the orbit. A given open orbit may lie preferentially in one or other hemisphere of the Fermi surface and therefore also have a z-component of motion and of field-independent effective path. In the plane normal to **B** extended motion is impossible at right angles to the open orbit, so that this component of **L** falls off as $1/\gamma^2$ just like a closed orbit. It is perhaps worth remarking that open orbit conductivity is achieved, as in other cases, by creating electrons on those orbits which have one sense of motion and destroying them on those of the opposite sense.

We may now write the general form of the conductivity tensor, analogous to (28), when open orbits are present. As usual, **B** lies along z, but we choose as x-axis the line of the open orbits in k-space, so that in real space the open orbits allow no motion along x. Then in high fields

$$\sigma_{ij}/\sigma_0 \approx \begin{bmatrix} a/\gamma^2 & d/\gamma & e/\gamma \\ -d/\gamma & \beta + b/\gamma^2 & \zeta \\ -e/\gamma & \zeta & c \end{bmatrix} \tag{3.31}$$

from which $\rho_{xx}/\rho_0 \approx (\beta c - \zeta^2)/\Delta$ and $\rho_{yy}/\rho_0 \approx (ac + e^2)/\gamma^2\Delta$. Since the determinant Δ has a leading term proportional to $1/\gamma^2, \rho_{xx}$ increases quadratically without limit while ρ_{yy} saturates.

It is only when **J** is made to flow exactly along the open-orbit direction in real space that this orbit can monopolize the current and allow the resistivity to saturate. For any other direction of **J** the open-orbit conductivity, wrongly directed, is a handicap to the other electrons. A vector diagram, fig. 19, illustrates this point;[26] only the xy-plane is considered, as if e and ζ were zero, and we assume the open-orbit

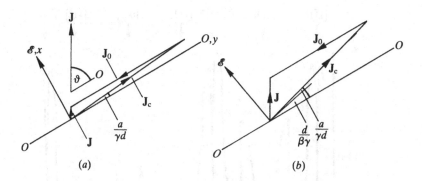

Figure 3.19 Vector diagrams illustrating effect of open orbits.

conductivity to be the largest contribution by far to σ_{ij}. Then to force the current along the prescribed (vertical) direction, when the open-orbit conductivity lies along the line O in (a), we might start by directing $\mathscr{E}$ normal to O so that no open-orbit current results. The other electrons, when $\gamma \gg 1$, generate a current $\mathbf{J_c}$ which is very nearly normal to $\mathscr{E}$ and not in the right direction. A very slight tilt of $\mathscr{E}$, however, brings the open orbits into play without significantly changing $\mathbf{J_c}$, and the new contribution $\mathbf{J_0}$ can be arranged to combine with $\mathbf{J_c}$ to give the required resultant $\mathbf{J}$. But $\mathbf{J}$ is now very small, and there is a substantial component of $\mathscr{E}$ parallel to it, so that the resistivity is large. The nearer the Hall angle is to $\pi/2$ for the closed-orbit electrons, the less there is left after $\mathbf{J_c}$ and $\mathbf{J_0}$ combine.

Quantitative refinement of the argument is best done by taking the x and y components of (31) and rotating the axes through $\pi/2 - \theta$ so that the x-axis coincides with the current flow. We also put $b = a$, assuming for convenience that the closed-orbit conductivity is isotropic. Then

$$\sigma_{ij}/\sigma_0 \approx \begin{pmatrix} \beta \cos^2\theta + a/\gamma^2 & \beta \sin\theta\cos\theta + d/\gamma \\ \beta \sin\theta\cos\theta - d/\gamma & \beta \sin^2\theta + a/\gamma^2 \end{pmatrix}. \tag{3.32}$$

Consequently,

$$\rho_{xx}/\rho_0 = \sigma_{yy}/\sigma_0\Delta^2 \approx \gamma^2\beta^2 \sin^2\theta/(a\beta + d^2), \tag{3.33}$$

showing the expected quadratic increase except when $\theta = 0$. If, as assumed in drawing fig. 19(a), the open orbit dominates the scene, to make $d^2 \ll a\beta$, (33) becomes $\rho_{xx}/\rho_0 \approx \gamma^2\beta \sin^2\theta$, which is the implication of the vector diagram as drawn. In reality, however, (see fig. 19(b)) a and d have magnitude something like unity, as in the free-electron expressions (1.16),

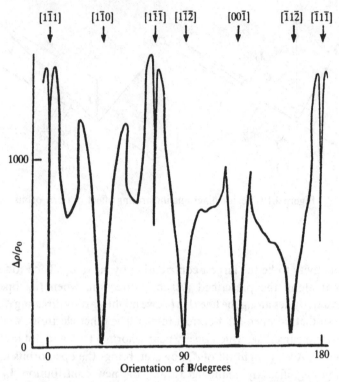

Figure 3.20 Variation of high-field transverse magnetoresistance in copper as **B** is turned through 180° in the (100) plane (Klauder *et al.*[22]). Principal symmetry directions, at which $\Delta\rho_{xx}/\rho_0$ falls to a low value, are shown by arrows.

while β represents the conductivity of a generally small fraction. Hence it is probably better to write

$$\underset{B\to\infty}{L}\,(\rho_{xx}/\rho_0)\simeq\gamma^2 f^2 \sin^2\theta, \qquad (3.34)$$

in which $f(=\beta/d)$ is the fraction of electrons involved in open orbits. As **B** is tilted away from a high-symmetry direction, e.g. [001] in Cu, f rises linearly with angle of tilt, and the quadratic term in ρ_{xx} as the square of the tilt. The difference between the saturation value at [001], $\rho_{xx}(\infty)$ being only a few times ρ_0, and the height of the quadratic rise which may easily be several powers of ten above ρ_0, leads to the dip of resistivity at the symmetry point being extremely sharp.

Fig. 20 is a rotation diagram for a copper matchstick sample with **J** along [110] and **B** in the plane normal to this; $\gamma \sim 70^*$. In the stereographic representation of fig. 17, **B** traverses the perimeter of the diagram, passing through the three principal symmetry directions and also through another direction [112], at which the magnetoresistance saturates. This is one of those special directions at which two separate sets of periodic orbits coexist (see fig. 17), as illustrated in fig. 21. The current, along [110], is normal to the paper, and typical periodic open orbits are shown. Thus an electron starting at A, on a neck, moves round to the back of the surface where it meets another neck immediately behind. The next replication, onto which it moves, joins it at A' so that the next stage is a longer arc of the surface to a point immediately behind A'; this in its turn is equivalent to the starting-point A, and the pattern repeats itself indefinitely. The general run of the k-orbit is parallel to **J**, so that the r-orbit is normal to **J**. If those orbits and their opposites, indicated by the line BB', were the only open orbits present, ρ_{xx} would rise indefinitely. There is also, however, the open orbit CC' running normal to the others, and the two sets of orbits together can carry

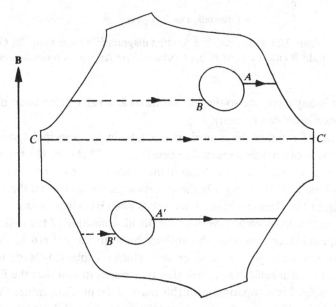

Figure 3.21 The two periodic open orbits in copper when **B** lies along [211].

---

* The minimum values of $\Delta\rho/\rho$ recorded here are not to be too strictly interpreted; the minima are so sharp (especially at [111]) that great care in alignment is needed to find them precisely, and this was not attempted in the work of these authors, who were more concerned with mapping the general features.

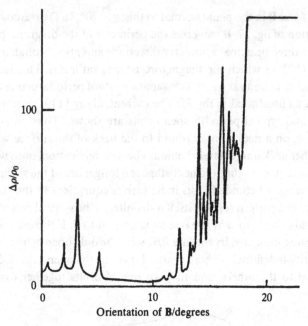

Figure 3.22 Fine detail in a rotation diagram[22] similar to fig. 20. On the right **B** enters a zone of open orbits where $\Delta\rho_{xx}/\rho_0$ is much higher.

current in any direction, so that ρ_{xx} saturates at a value determined by the total open-orbit conductivity.

When the rotation diagram does not take in the principal symmetry directions it can develop much fine detail, as fig. 22 shows. On the right, where the curve goes off-scale, **B** lies in the shaded region of aperiodic open orbits around [100] (see fig. 17); the spikes on the left arise from the spikes in the figure that denote periodic open orbits when **B** lies in a commensurate direction. It is from such traverses, and the identification of the indices of these special directions, that the authors built up the pattern in fig. 17. There is no need to go into detail on this, which is explained clearly in the paper, and it will suffice to remark that they were satisfied that the Fermi surface in fig. 5 is compatible with this mass of detail. This, rather than a critical study of magnetoresistivity, was the motivation of the work, which comprises one of the most thorough applications of the general ideas of LAK theory to the determination of Fermi surface shape. Another example, making use of the rotating sphere technique, will be found in the work of Dixon, from which fig. 2.15 was taken.

The technique of rotating a sample in a steady field was originally

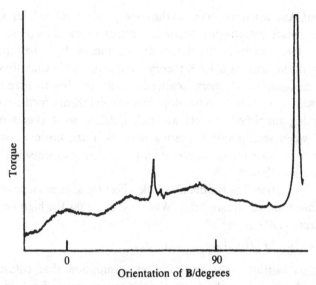

Figure 3.23 Torque on a single crystal of mercury as **B** is rotated evenly in a plane normal to the axis of suspension (Moss and Datars[27]).

devised as an easy way of plotting open-orbit directions,[27] and fig. 23 shows one of the first applications, to mercury. It is not clear what shape of sample was used, but it was not a sphere; for diagnostic purposes, however, this is irrelevant. As already remarked the method, unlike matchstick measurements, distinguishes between compensation and open orbits, both of which give non-saturating resistivity in a matchstick. Fig. 23 shows that the compensated metal mercury, for which non-saturating resistivity can be expected for almost every direction of **B**, has an open orbit when **B** is appropriately directed; and the position is very precisely indicated by the spike of torque.

Quantitative aspects of high-field magnetoresistance

1. Closed orbits

We noted in chapter 1 that in strong fields the Hall conductivity σ_{yx} does not depend on the details of scattering. This is hardly surprising since σ_{yx} becomes independent of τ. It is tempting to assume that σ_{xx} and other coefficients that vary as τ^{-1} may also in the limit be determined by some simple average of τ, but this hope is not realized. To be sure, the details of $\tau(\mathbf{k})$ are not as important as the shape of the Fermi surface, but they cannot

be ignored. And when we abandon the concept of a local $\tau(\mathbf{k})$, for example
when we replace catastrophic scattering events by small-angle scattering
processes, the changes in the theory are significant. It is the latter that
exemplify the limitations of LAK theory – not in principle, since they wrote
expressions containing general scattering processes, but in practice. No
attempt will be made here to develop their general theory further; instead,
by analysing simplified models we shall indicate what effects may be
expected when small-angle scattering prevails. First, however, we shall
continue to assume the existence of $\tau(\mathbf{k})$ and derive expressions for the
conductivity in these terms.

As an illustration that the details of $\tau(\mathbf{k})$ affect the answer let us return to
the two-dimensional square orbit. When τ is constant the high-field limit
follows from (1.86) and (1.87):

$$\sigma_{xx} \approx \pi^2 J_{xo}/12\omega_c^2\tau, \quad \sigma_{yx} \approx J_{xo}/4\omega_c. \tag{3.35}$$

Now let us instead make the extreme assumption that catastrophic
scattering occurs only at the corners of the square, with a fraction $\pi/2\omega_c\tau$
disappearing at each corner; the mean rate of decay is the same as before.
Unit impulsive field applied parallel to a side of the square generates J_{xo}
which subsequently varies with time to give a triangular waveform, as in
fig. 1.20, if there is no decrement. Scattering at the corners, however, lowers
the height of each triangle by a factor $C = (1 - \pi/2\omega_c\tau)^2$ compared to the
preceding triangle. The geometric series of triangular areas is easily
summed to give

$$\sigma_{xx} = \pi J_{xo}(1 - C)/4\omega_c(1 + C) \approx \pi^2 J_{xo}/8\omega_c^2\tau, \tag{3.36}$$

while σ_{yx} is unchanged. Hence $\sigma_{xx}(\infty)$ and consequently $\rho_{xx}(\infty)$ are
increased by a factor 3/2 when all the scattering is concentrated at the
corners.

In deriving the high-field limit of σ_{yx}, as in (1.49), it was possible to
assume that the terminal point on the k-plane was, in the limit, the same for
all electrons in a given orbit, irrespective of their starting point. This is not
good enough for σ_{xx}, as can be seen by writing the contribution of a slice δk_z.
Let an electron starting from a point s on the k-orbit, with coordinate $\mathbf{k}(s)$,
have as its mean terminal point $\mathbf{k}^T(s)$. Then the x-component, L_x, of its
effective path is $|\hbar/eB|[k_y^T(s) - k_y^{(s)}]$, and from (1.37),

$$\delta\sigma_{xx} = (e\delta k_z/4\pi^3 B) \oint (k_y^T - k_y)(dk_y/ds)\,ds. \tag{3.37}$$

The second term in the brackets vanishes identically on integrating round
the orbit, as does the first if k_y^T is independent of s. The consequent necessity

of determining the variation of k_y^T round the orbit makes the calculation of σ_{xx} considerably longer than for σ_{yx}, but the result in the end is not very complicated; if $l(\mathbf{k})$ is the mean free path for a Fermi electron, the high-field limit of σ_{xx} is expressed in terms of integrals over the Fermi surface:

$$\sigma_{xx} \approx (\hbar/4\pi^3 B^2) \left\{ \int_S k_y^2 \, dS/l - \left[\int_S k_y \, dS/l \right]^2 \Big/ \int_S dS/l \right\}. \quad (3.38)$$

This expression has the same form as that for the component I_{yy} of the inertia tensor (second moment) of a solid body, dS/l being interpreted as an element of mass. It is independent of choice of origin, but if the centroid of the Fermi surface, weighted in proportion to $1/l$, is chosen as origin, the second term in (38) vanishes, leaving a simple integral to be evaluated. Clearly it is advantageous, if the Fermi surface is reasonably uncomplicated and centro-symmetric, to choose its geometric centroid as origin, whereupon the second term is automatically eliminated. The result expressed in (38) was derived in a somewhat different form by Chambers,[28] who attributes it in the first place to Kohler.[29] The derivation that follows, though still lengthier than one would like, is more straightforward than either of its predecessors.

[We follow the fortunes of a bunch of N_0 electrons starting from s_0 on a chosen k-orbit. Integrating (12) we find that $\delta N(s)$ of these are scattered catastrophically, and effectively brought to rest, in an element δs at s, where

$$\delta N(s) = (N_0/\alpha l') \exp\left(-\int_{s_0}^{s} ds/\alpha l' \right) \delta s \approx (N_0/\alpha l') \left[1 - \int_{s_0}^{s} ds/\alpha l' \right] \delta s, \quad (3.39)$$

in which l' is written for $l \sin \theta$, the projection of $\mathbf{l}$ onto the plane normal to $\mathbf{B}$. The centroid of all electrons brought to rest defines the terminal point, whose y-component, k_y^T, is thus given by

$$k_y^T(s_0) = \oint k_y(dN/ds)\, ds \Big/ \oint (dN/ds)\, ds. \quad (3.40)$$

There is no need to integrate round more than one orbit, starting from s_0, since all subsequent orbits by surviving electrons contribute the same as (40). When substituting for dN/ds from (39) we observe that the first term in the square brackets gives a term in k_y^T that is independent of s_0 and may therefore be ignored. We then have

$$k_y^T(s_0) \approx -\alpha^{-1} \oint (k_y \, ds/l') \int_{s_0}^{s} ds/l' \Big/ \oint ds/l'. \quad (3.41)$$

The integral $\int_{s_0}^{s} ds/l'$ must be taken round the orbit in the direction of electron motion. If s' is reached from s_0 after passing through the origin of s (which may be arbitrarily chosen),

$$\int_{s_0}^{s'} ds/l' = \oint ds/l' + \int_0^{s'} ds/l' - \int_0^{s_0} ds/l',$$

but if the origin is not crossed the first term is absent. With this in mind we rewrite (41):

$$k_y^T(s_0) \approx -\alpha^{-1} \left\{ \int_0^{s_0} k_y ds/l' + \oint (k_y ds/l') \right.$$

$$\left. \times \left[\int_0^s ds/l' - \int_0^{s_0} ds/l' \right] \middle/ \oint ds/l' \right\}. \tag{3.42}$$

The first term in the square brackets does not contain s_0 and may be ignored; the second is independent of the line-integral outside the brackets. On substituting (42) in (37), we find:

$$\delta\sigma_{xx} \approx -(\hbar\delta k_z/4\pi^3 B^2) \left\{ \oint (dk_y/ds_0) ds_0 \int_0^{s_0} k_y ds/l' \right.$$

$$\left. - \oint (dk_y/ds_0) ds_0 \int_0^{s_0} ds/l' \times \oint k_y ds/l' \middle/ \oint ds/l' \right\}. \tag{3.43}$$

The two double integrals may be simplified by reversing the order of integration (with appropriate care as to the limits of integration). Thus:

$$\oint (dk_y/ds_0) ds_0 \int_0^{s_0} k_y ds/l' = \oint (dk_y/ds_0) ds_0 \times \oint k_y ds/l'$$

$$- \oint (k_y ds/l') \int_0^s (dk_y/ds_0) ds_0$$

$$= -\oint k_y^2 ds/l' + k_y(0) \oint k_y ds/l'. \tag{3.44}$$

And

$$\oint (dk_y/ds_0) ds_0 \int_0^{s_0} ds/l' = \oint (dk_y/ds_0) ds_0 \times \oint ds/l'$$

$$- \oint (ds/l') \int_0^s (dk_y/ds_0) ds_0$$

$$= -\oint k_y ds/l' + k_y(0) \oint ds/l'. \tag{3.45}$$

Inserting (44) and (45) in (43) gives:

$$\delta\sigma_{xx} \approx (\hbar\delta k_z/4\pi^3 B^2) \left\{ \oint k_y^2 ds/l' - \left[\oint k_y ds/l' \right]^2 \middle/ \oint ds/l' \right\}. \tag{3.46}$$

Finally, we put $\delta k_z ds/l' = dS/l$, and integrate over all slices to reach (38).]

2. Extended orbits

The result expressed by (38) is well illustrated by an extended orbit such as is shown in fig. 16(*a*). This particular example has no 'parasitic' units, by which is meant units that could be cut off without shortening the chain; the units labelled *P* are parasitic to the open orbit of fig. 16(*b*). Of course, they extend the path an electron must take and slow down its mean velocity along the side of the orbit. Another example is the rather short extended orbit of fig. 24(*a*), in which one of the seven units is parasitic, breaking what would otherwise be dyad symmetry, as in fig. 16(*a*). In longer orbits the sinuosities and parasites tend to be evenly distributed, so that the position of the centroid is found by inspection with sufficient precision to avoid the necessity of keeping the second term in (38). One may also note that neck and belly regions of the Fermi surface are closely intermixed, so that even if *l* varies considerably between the two it is likely that a mean value will suffice to describe the scattering in high fields, and that it may be assumed constant along the length of the orbit. In view of the form of (38), the appropriate

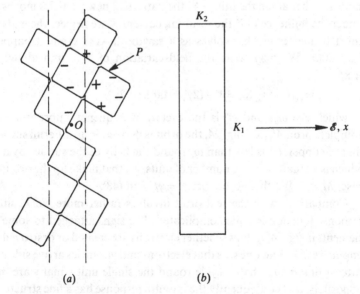

(*a*)   (*b*)

Figure 3.24 (*a*) An asymmetrical extended orbit with its centroid at *O* and a single parasitic orbit *P*; the signs indicate whether electrons are created or destroyed by application of $\mathscr{E}$ in the direction shown. (*b*) the equivalent rectangular orbit.

mean is

$$\bar{l} = \langle 1/l \rangle^{-1}, \tag{3.47}$$

the brackets indicating an average over those areas of the Fermi surface in a slice δk_z that gives rise to the orbit in question.

It is now convenient to define the sinuosity ξ as the ratio of the actual length of the orbit, e.g. fig. 24(*a*) to the perimeter of a rectangle (*b*) that matches the real orbit in overall length K_1, and has a width K_2, such that $K_1 K_2$ equals the area of the real orbit. We shall now compare the contributions of the two to the high-field conductivity, noting that the equality of areas ensures that each has the same limiting form of $\delta\sigma_{yx}$. As for $\delta\sigma_{xx}$, in this case the direction of high conductivity with $\mathscr{E}$ as shown in the diagram, the only difference is that dS in (38) must be rewritten as ξdS for the real orbit. In other words, if we interpret $1/l$ as $1/\bar{l}$ and multiply the result by ξ, the rectangular replacement will give the same high-field limit as the real extended orbit.

To look at this in more detail, let us compare the two, considered as two-dimensional systems, over the whole range of field strength. First, the rectangular orbit; the current response to an impulsive field $\mathscr{E}_x$ arises from the generation of electrons evenly along one side and the elimination of the same number along the other. As the pattern of new particles moves round under the influence of $\mathbf{B}$ the resulting current, if we neglect the ends as too short to matter much, evolves as a sawtooth – the same, it happens, as fig. 1.20(*a*). We may write the field-variation of $\delta\sigma_{xx}$ immediately from (1.86):

$$\delta\sigma_{xx}(B)/\delta\sigma_{xx}(0) = 1 - (2\gamma'/\pi)\tanh(\pi/2\gamma'), \tag{3.48}$$

in which $\gamma' = \omega_c'\tau$ and ω_c' is the cyclotron frequency appropriate to the complete orbit. If, as in fig. 24, the orbit is derived from a Fermi surface like that of copper, ω_c' is less than ω_c round the belly of the surface by a factor which is virtually just the number of units, M, that make up the orbit; in this case, $M = 7$. We therefore write $\gamma' = \gamma/M$ in (48).

Comparison with the real orbit involves rather more work, since the response function is quite complicated. The signs attached to some of the elements in fig. 24(*a*) show whether electrons are created or destroyed by the impulsive field, and one sees that electrons and vacancies arrive side by side. After a time π/ω_c – half a cycle round the single unit – many are moving oppositely, and consequently the impulse response has a fine structure that is lacking in that of the rectangular orbit. The two responses are shown in fig. 25, which makes clear that the fine structure is a high-frequency ripple on the overall sawtooth. When $\gamma \gg 1$ this ripple plays little part, its influence

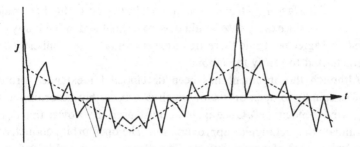

Figure 3.25 Impulse response of J_x for the orbit in fig. 24(a), shown by a full line, and (broken line) for the equivalent rectangular orbit (b).

on $\delta\sigma_{xx}$ having decayed as $1/\gamma^2$ just as with a small closed orbit. On the other hand, when $\gamma \ll 1$ most electrons are scattered while still on the first spike of the response function, where the difference between the real response and the sawtooth is considerable. This means no more than that it is irrelevant in low fields whether the orbit runs across necks or not.

To calculate the conductivity the response function must be multiplied by $\exp(-t/\tau)$ and integrated. The result is displayed in fig. 26, which shows the ratio of the actual $\delta\sigma_{xx}$ to that predicted by (48). There is agreement on the left, where $\gamma \gg 1$, and it is interesting to note that when γ is as low as 2 the difference is only about 20%. For many purposes, then, the rectangular model is very adequate. The greatest discrepancy, when $\mathbf{B} = 0$, amounts to a factor 2.4 in this case, which may be compared with ζ which is 2.5 if the ends of the rectangles are disregarded, 2.0 if they are not. Exact agreement is not to be expected, but the general conclusion is clear – smoothing out the sinuosity is admissible when $\omega_c\tau$ is large, but entails an error something like

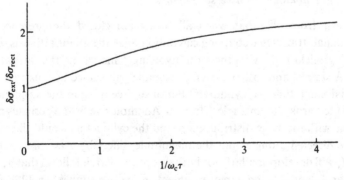

Figure 3.26 Comparison of $\delta\sigma_{xx}$ for the two orbits in fig. 24. Note that the ratio is shown as a function of $1/\omega_c\tau$.

the smoothing factor ζ when it is small. To take the matter further would be unprofitable, since every case would have to be analysed individually to get numerical agreement, and there are no experimental results sufficiently well characterized to justify the labour.

Although the argument has been developed for extended orbits, it applies equally to open orbits for which $\omega' = 0$, so that according to (48), $\delta\sigma_{xx}$ is independent of **B**. Once again we must insert the proviso that $\omega_c\tau > 2$ to validate the rectangular approximation. The open orbit conductivity is the same as that of a strip with straight edges, but reduced by a factor ζ.

This part of the discussion, deriving the conductivity due to those electrons that move in extended or open orbits, is the easy part of any calculation aimed at comparing theory and experiment. As the discussion around fig. 17 indicates, the fraction of electrons involved in such orbits, and even more so the length distribution of the extended orbits, varies violently with field orientation, creating serious problems of enumeration. In one attempt[30] to handle the problem, as a part of a calculation of magnetoresistance in polycrystalline copper, a statistical approximation was made. The underlying idea was that the electrons probably suffer small deflections by encounters with phonons, dislocations, etc., insufficient to affect the low-field conductivity significantly but sufficient to provide, as it were, a shimmer on the broken lines in figs. 16 and 24. Instead of a well-defined orbit the cutting of a neck now becomes a matter of chance, and an electron starting with a given **k** may exhibit a statistical ensemble of possible orbit lengths. Details can be found in the original paper; the idea remains untested except by this single, hardly critical, application and cannot be regarded as an established procedure.

3. Longitudinal–transverse coupling

It is clear from (29) that when all orbits are closed the presence of longitudinal–transverse coupling can only change the quantitative aspects of the high-field resistivity, the term involving f increasing the saturation value. A simple, and rather extreme, example will suffice to illustrate the potential magnitude – a cylindrical Fermi surface lying in the xz-plane is inclined towards x at an angle ψ from **B**. An impulsive field $\mathscr{E}_z$ creates new particles with the same distribution round the cylinder as would the same impulse directed along x, but the magnitude differs by a factor $-\tan\psi$. Since J_x will develop similarly for both directions of $\mathscr{E}$, it follows that $\sigma_{xz} = -\sigma_{xx}\tan\psi$; and by the same argument, $\sigma_{yz} = -\sigma_{yx}\tan\psi$. In (28) $f = d\tan\psi$, and if b, c and d are comparable the resistivity is multiplied by

$\sec^2 \psi$ as a result of not ignoring the coupling. Usually, such tilted arms are accompanied by other parts of the Fermi surface which dilute their influence. If they are not, as in conductors (graphite is perhaps the nearest example) with a dominant cylindrical surface, the conductivity is almost perfectly planar, being very small along the axis of the cylinder, and the appearance of longitudinal–transverse coupling is not surprising. It is there in zero field to the same degree.

As for open orbits, the periodic orbits labelled AA' and BB' in fig. 21 can contribute a field-independent coupling term at high fields by virtue of their asymmetry. A field $\mathscr{E}_x$, in the plane of the paper, excites as many electrons on the short section A as on the longer section A', since both have the same extension normal to the paper. Those on A move upwards and those on A' downwards, so that the resultant J_z arises from the different z-velocities and the different times spent on the two sectors. On the other hand all move along the open orbit together and, even after allowing for sinuosity, one can expect the resultant J_x to be at least twice J_z, and perhaps considerably more. All the same, σ_{zx} can have a non-negligible high-field limit, but these two open orbits do not contribute a field-free term to σ_{zy}. Nor does the orbit CC' or, more precisely, the set of similar orbits; for every one above the equator, with a net upward velocity component, there is one below, equally excited by $\mathscr{E}_y$, moving downwards to neutralize J_z. It follows from (31) that the transverse–longitudinal coupling in the open orbits does not change the qualitative behaviour of ρ_{xx} and ρ_{yy}, and it seems likely that the quantitative change is not very pronounced. Whether there are open orbits or not, we cannot expect to find anything relating to longitudinal–transverse coupling except from careful measurements and extensive computation on the basis of a known Fermi surface and variation of l over it.

4. Longitudinal magnetoresistance

In the framework of LAK theory, with catastrophic scattering, the longitudinal magnetoresistance always saturates at high fields; in the absence of open orbits, (28) shows that $\rho_{zz}/\rho_0 \approx 1/c$, while if open orbits are present, as described by (31), the value $1/c$ must be corrected, probably by only a little. It is easy to derive a formal expression for the limit of c; the care needed for σ_{xx}, varying as $1/B^2$, and even the modest care needed for σ_{yx}, varying as $1/B$, are here unnecessary when there is no field-variation. As $\omega_c\tau \to \infty$, every electron on a given orbit may be assumed to traverse so many cycles that all have the same distribution of path lengths in the z-direction. From (39) we write the probability of collision in one cycle of the helical path as $\alpha^{-1}\oint ds/l'$, and the distance travelled is given by (1.30); these two results allow us to write the effective path immediately as

$$L_z = (\partial \mathscr{A}_k/\partial k_z)_E \Big/ \oint ds/l'. \tag{3.49}$$

Now the Fermi surface contained in a slice δk_z has projected area on a plane normal to **B** given by:

$$\delta S_z = (\partial \mathscr{A}_k/\partial k_z)_E \delta k_z, \tag{3.50}$$

and these two expressions may be substituted in (1.37) to give

$$\sigma_{zz}(\infty) = (e^2/4\pi^3\hbar) \int dk_z (\partial \mathscr{A}_k/\partial k_z)_E^2 \Big/ \oint ds/l'. \tag{3.51}$$

This may be simplified in form by noting that $dS = ds\, dk_z \operatorname{cosec} \theta$, so that

$$\sigma_{zz}(\infty) = (e^2/4\pi^3\hbar) \int \bar{l} C^2 \, dS, \tag{3.52}$$

$\bar{l}$ being defined as in (47). The quantities C and $\bar{l}$, though assigned to each point on the Fermi surface, are in fact functions of k_z alone. Thus C is written for $(\partial \mathscr{A}_k/\partial S)_E$, and since $\mathscr{A}_k$ and S_z are identical for a closed surface, C is a certain mean of $\cos \theta$ taken round a k-orbit. And $\bar{l}$ is an orbit average.

For the zero-field behaviour (1.37) gives:

$$\sigma_{zz}(0) = (e^2/4\pi^3\hbar) \int \cos^2 \theta \, dS, \tag{3.53}$$

so that

$$c = \sigma_{zz}(\infty)/\sigma_{zz}(0) = \int l \cos^2 \theta \, dS \Big/ \int \bar{l} C^2 \, dS. \tag{3.54}$$

If l is constant, it cancels with $\bar{l}$ to leave a purely geometrical expression,

$$c = \int \cos^2 \theta \, dS \Big/ \int C^2 \, dS, \tag{3.55}$$

and since $C = \cos \theta$ for a spherical surface, or any surface with axial symmetry about **B**, $c = 1$ in such cases, and there is no longitudinal magnetoresistance (as is quite obvious). The argument has been presented for closed Fermi surfaces, but can be extended to open surfaces if necessary.

Powell[31] has computed c for copper from the measured Fermi surface and with a variety of hypothetical variations of l. For the [100] orientation he finds constant l gives $\rho_{zz}(\infty)/\rho_{zz}(0)$ to be close to 2, while three samples with different resistance ratios gave 2.0, 1.9 and 1.7. The lower values are consistent with l being markedly shorter near the neck regions, which is plausible. There is no reason why different samples should not show different patterns of scattering; also, as we shall see presently, the notion of catastrophic scattering is often a very imperfect assumption.

A much simplified model of the Fermi surface of copper, not unlike that which was implicit in the derivation of the curve in fig. 18, serves to estimate $\rho_{zz}(\infty)/\rho_{zz}(0)$ and will prove useful later. The real surface is replaced by a sphere and the necks shortened so that the unit of the replicated surface is as shown in fig. 27. In giving the necks the same angular diameter, 20°, as on the real surface their influence is undoubtedly underestimated, but we shall not be so much concerned with quantitative agreement as with providing a basis for making rough estimates when we come to abandon the assumption of catastrophic scattering. Still adhering to it, we note that when $\omega_c\tau \gg 1$ electrons excited by $\mathscr{E}$ in a section that cuts a neck are likely to move on an orbit of small pitch. Thus if **B** is oriented so that only one neck is cut in the centre, the path on the replicated surface is a closed orbit

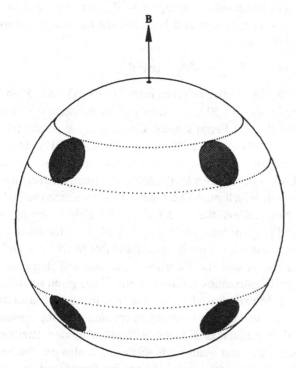

Figure 3.27 Simplified model of the Fermi surface of copper, with the neck zones when **B** ∥ [100] shown by dotted lines.

around two units, on which v_z is oppositely directed. Such an electron, for which $\bar{v}_z = 0$, makes no high-field contribution to J_z. The same is true for orbits which involve 3 or more necks, as when **B** is along [111] for 6-neck hole orbits, or [100] for 4-neck hole orbits. Only when 2 necks are involved, as in fig. 21, will the electrons make a (relatively minor) contribution to J_z. Around [225] a substantial fraction of the electrons are eliminated in this way, and quite a high ratio $\Delta\rho(\infty)/\rho_0$ is to be expected. Ström-Olsen[32], who worked with silver and used the same model but with a different neck radius, calculated that for **B** along [111], where the necks have least effect, $\Delta\rho(\infty)/\rho_0$ should be only 0.094, while along [147], which is not the worst case, it should be 1.77. With samples whose impurity content might be expected to give predominantly large-angle, i.e. catastrophic, scattering he found the experimental ratios to be rather larger but not uncomfortably so in view of the model used.

In this model, with l taken as constant, C in (52) is $\cos\theta$ and dS is $\sin\theta\,d\theta$ (constants are neglected) and the integral is to be taken only over such values of θ where no necks intercept the orbit. Thus any neck-free belt between θ_1 and θ_2 contributes $|\cos^3\theta_1 - \cos^3\theta_2|$ to $\sigma_{zz}(\infty)$. If it were not for the necks, this term in σ_0 would be 2, but the necks are estimated to reduce this to 1.88. Hence:

$$\rho_{zz}(\infty)/\rho_0 = 1.88/\sum|\cos^3\theta_1 - \cos^3\theta_2|. \qquad (3.56)$$

For **B** along [100] the effective belts run from 0 to 45°, 65° to 115° and 135° to 180°, and $\rho_{zz}(\infty)/\rho_0 = 1.30$ (1.98); the figure in brackets is Powell's[31] computation for the real Fermi surface. Corresponding results for other orientations are: [110] – 1.56 (2.59) and [111] – 1.13 (1.17). The major cause of the discrepancy is the shape of the real surface. Though not very far from spherical, apart from the necks, the difference greatly reduces the pitch of the helical orbits which might be expected to dominate the high-field conductivity. Nevertheless the model gives the right ordering of the behaviour, with [111] the least affected by **B**, and [110] the most. Ström-Olsen's better success with silver is understandable in that the necks are smaller than in copper and the distortion from spherical shape less.

A similar type of calculation for aluminium[33] has given a satisfactory account of the field-variation of ρ_{zz} except that instead of saturating it follows in practice the common tendency to continue drifting upwards, in contradiction of the simplest LAK theory with catastrophic scattering. One possible explanation is that small-angle scattering is also present, and the model of fig. 26 now provides a suitable basis for demonstrating how this can radically alter the picture.

Small-angle scattering

The most evident source of small-angle scattering at low temperatures is the thermal excitation of the lattice. Phonons of energy around $3 k_B T$ are most effective, with wave-vectors about $3 T/\Theta_D$ times the radius of a Brillouin zone. In copper k_F is also comparable to the radius of the zone, and $\Theta_D \sim 300$ K, so that in absorbing or creating a phonon at, say, 4 K the electron can be scattered through no more than a few degrees, distinctly less than the width of a neck. The electron performs a random walk over the Fermi surface, its mean square displacement being proportional to time. For such processes, acting alone, to destroy an organized current on a spherical Fermi surface, the electrons must have time to wander through about $\pi/2$. To clarify the argument, let us assume the typical scattering angle to be so small that the electron distribution on the surface obeys a diffusion equation.

With $\mathscr{E}_z$ applied steadily the equilibrium distribution of extra electrons on the surface of the sphere will be axially symmetric, with a number density $n(\theta)$ per unit area of Fermi surface. The flux of particles across a line of latitude is $- 2\pi D \sin \theta \cdot \mathrm{d}n/\mathrm{d}\theta$, D being the diffusion coefficient, so that a belt $\delta\theta$ wide gains particles at a rate $2\pi D\,(\mathrm{d}/\mathrm{d}\theta)(\sin \theta\, \mathrm{d}n/\mathrm{d}\theta)\,\delta\theta$. At the same time $\mathscr{E}_z$ increases the number at a rate $(k_F^2 e\mathscr{E}_z/2\pi^2\hbar)\sin \theta \cos \theta\, \delta\theta$, so that in the steady state

$$(\mathrm{d}/\mathrm{d}\theta)(\sin \theta\, \mathrm{d}n/\mathrm{d}\theta) + 2p \sin \theta \cos \theta = 0, \qquad (3.57)$$

where $p = k_F^2 e\mathscr{E}_z/8\pi^3\hbar D$. This is the fundamental equation for the problems that follow. Symmetry demands that $n(\pi/2) = 0$ always and $n(\theta) = -n(\pi - \theta)$.

When there are no necks the solution of (57) is

$$n = p \cos \theta. \qquad (3.58)$$

The same distribution obtains when scattering is catastrophic, with constant relaxation time τ_c, when the increase due to $\mathscr{E}_z$ is balanced by a scattering rate $2\pi k_F^2 n \sin \theta\, \delta\theta/\tau_c$ and

$$n = e\mathscr{E}_z\tau_c \cos \theta/4\pi^3\hbar = (2D\tau_c/k_F^2)\, p \cos \theta. \qquad (3.59)$$

The relaxation time needed to make (58) and (59) agree is $k_F^2/2D$ which, confirming the remark above, is the time taken in a random walk on a sphere for a root mean square angular displacement of about $\pi/2$. We write $k_F^2/2D$ as τ_d.

A more instructive way of looking at this result,[33] from the point of view of longitudinal magnetoresistance in the model of fig. 27, is to note that an

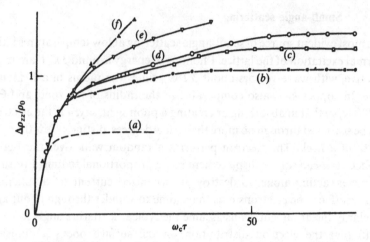

Figure 3.28 Longitudinal magnetoresistance of copper (Ström–Olsen[32]); at 4.2 K, $r_0 = 1.2 \times 10^4$; **B** and **J** lie 6° from [100] towards [010]. (*a*) theoretical for catastrophic scattering (*b*) measured at 4.2 K (*c*) at 7 K (*d*) at 10 K (*e*) at 13 K and (*f*) at 15 K.

electron in one of the neck zones is bound to be swept into a neck when **B** lies along [100], and lost to the conduction process, if it cannot escape within a quarter of a cycle, or $\pi/2\omega_c$. It is when this condition is satisfied that the resistivity approaches saturation. With catastrophic scattering this implies that the approach is well advanced when $\omega_c\tau$ reaches $\pi/2$. With diffusive motion, however, the time t taken to move 10° (0.17 radians), half the width of the zone, is given by $(0.17\,k_F)^2 = 2Dt$, or $t = \tau_d/33$. In a conventional plot of $\Delta\rho/\rho_0$ against $\omega_c\tau$, saturation would still be incomplete when $\omega_c\tau = 33\,\pi/2 = 52$. In the light of this one can appreciate in general terms why raising the temperature, and hence the phonon scattering, results in a greater longitudinal magnetoresistance in copper, and a delay in saturation, as shown by fig. 28*. Even at 4 K, where the phonon component is weak, there is evidence of considerable small-angle scattering – especially if the samples have been treated to improve their low-temperature conductivity. In copper the residual resistance is usually caused by scattering from small amounts of transition metal impurity (e.g. Mn), which can be made ineffective by prolonged heating in oxygen at a low

---

* Though not relevant to this aspect of the problem, the sharp rise at low values of $\omega_c\tau$ deserves notice. As with the transverse magnetoresistance, the small size of the neck is responsible; it requires only a short fraction of the orbit period for v_z to be reversed as an electron passes across a neck. If the reversal were instantaneous there would be no quadratic term in the low-field magnetoresistance.

pressure. It is believed that the impurity is oxidized and then precipitates as small inclusions whose overall scattering power is greatly diminished. What is left, however, is likely to have a marked small-angle scattering lobe, simply by virtue of its greater size. In consequence, untreated copper usually has a poor residual resistance ratio (~ 1000 at best) but shows magnetoresistance conforming fairly well to catastrophic scattering, while treated copper may have r_0 as high as $50\,000$ but suffers from great uncertainty as to the nature of the scattering.

At the extreme opposite to catastrophic scattering we have purely diffusive motion on the Fermi surface, as described by (57). At intermediate field strengths, when an electron may be swept by **B** into a neck, or may escape from the neck zone before suffering reversal of v_z, the theory is very difficult. In the high-field limit, however, every electron straying into a neck zone undergoes so many reversals that it has an equal chance of emerging in the top or bottom hemisphere. Then on the average v_z is destroyed and the act of entering a neck zone is equivalent to destruction. The boundary condition to be imposed on (57) is then that $n = 0$ at the edges of the zone. The general solution is

$$n = p(\cos \theta + C_1 + C_2 \ln \tan \tfrac{1}{2}\theta), \qquad (3.60)$$

and the vanishing of n at θ_0 and θ_1, the edges of the neck zone in the upper hemisphere ($\theta_1 > \theta_0$), determines the constants C_1 and C_2, and hence J_z. If we write σ_f for the free-electron conductivity derived from (58), we find:

$$\sigma_{zz}(\infty)/\sigma_f = \tfrac{1}{2}(1 - \cos \theta_0)^2(2 + \cos \theta_0) + \tfrac{1}{2}\cos \theta_1(3 - \cos^2 \theta_1)$$
$$+ 3\cos^2 \theta_1/2 \ln \tan \tfrac{1}{2}\theta_1. \qquad (3.61)$$

With

$$\theta_0 = 44.7° \quad \text{and} \quad \theta_1 = 64.7°, \quad \rho_{zz}(\infty)/\rho_f = 8.66, \qquad (3.62)$$

We also need to know the zero-field resistivity, which in the diffusion limit is more seriously affected by the presence of the necks than one might guess. On the replicated model there is a point near the centre of each neck at which symmetry demands n would vanish if the point were actually on the Fermi surface; one may then be confident that n is very small on the neck regions of the surface. In order for its path to be effectively terminated an electron need not diffuse to the equatorial belt but may reach a neck, after which it is equally likely to have positive or negative v_z. And since the time taken to travel a given distance varies as the square of the distance, the necks, small though they are, are rather efficient sinks by virtue of being evenly distributed on the sphere. To solve the diffusion equation, supplemented by particle creation, when $\mathscr{E}$ is present, approximate methods are needed, of which that used by Klemens and Jackson[34] is a good example. They point out that in the absence of $\mathscr{E}$ one may write a solution of the diffusion equation when only two necks are present at opposite ends of a diameter,

$$n = C_2 \ln \tan \tfrac{1}{2}\theta', \tag{3.63}$$

θ' being measured from the diameter joining the necks. Four such expressions, with the same C_2 but with θ' measured with respect to the four diameters joining pairs of opposite necks, add up to a solution which is only singular at the centres of the necks, where the solution is not needed. Otherwise it satisfies the diffusion equation. Finally they add a term $p \cos \theta$ to represent the creation of particles by $\mathscr{E}$, and their diffusion-controlled equilibrium distribution; they then adjust C_2 so that the mean value of n vanishes at the rim of the necks. This recipe provides a genuine solution of (57) at every point outside the necks, with n so low on the rims that it must be a very near approximation to the correct answer. With necks of angular radius $10°$, σ_0/σ_f turns out to be 0.345 when $\mathscr{E}$ is chosen to lie along [100] or [110], and 0.352 when [111] is chosen. Of course, these should be the same, and the difference is an indication of the error arising from the approximation – quite negligible in this case. Taking 0.347 as a mean, and combining it with (62), we have for the [100] direction,

$$\rho_{zz}(\infty)/\rho_0 = 3.0,$$

or $\Delta\rho_{zz}(\infty)/\rho_0 = 2.0$ instead of 0.30 as found for catastrophic scattering.

This estimate is more in line with curve (f) of fig. 28, though the rough nature of the model does not justify drawing firm conclusions. It is, however, worth examining an intermediate case – diffusion and catastrophic scattering both present – to see whether the temperature variation shown in the diagram can be accounted for in this way; the modified form of (57) may now be written:

$$(\mathrm{d}/\mathrm{d}\theta)(\sin\theta \, \mathrm{d}n/\mathrm{d}\theta) + 2\sin\theta(p\cos\theta - cn) = 0, \tag{3.64}$$

in which $c = k_F^2/2\tau_c D = \tau_d/\tau_c$. For the free-electron sphere the solution is $n = p\cos\theta/(1 + c)$, so that c does indeed measure the relative importance of catastrophic and diffusive processes for the free-electron resistivity. We have seen, however, that the necks enhance the diffusive resistivity, in zero field, by a factor 2.88 and the catastrophic by 1.06; the relative importance of catastrophic processes in the model is therefore better measured by $c/2.72$.

Analysis of (64) is troublesome, but computation easy, in the high-field limit. At the edge of a neck zone, and on the equator, where $n = 0$, one can

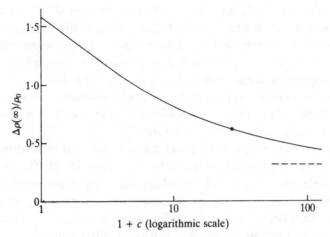

Figure 3.29 Saturation value of longitudinal magnetoresistance for the model of fig. 27, with $\mathbf{B} \parallel [100]$; c is a measure of the importance of catastrophic scattering relative to diffusion.

choose an initial value of $dn/d\theta$ and integrate step-by-step until another neck zone or the pole is reached. In the former case the starting condition is adjusted to make n vanish at the second edge; in the latter to make $dn/d\theta$ vanish at the pole. The result is shown in fig. 29 for B along [100], and it is very clear that small amounts of small-angle scattering can greatly enhance the high-field limit. Thus at the point marked on the curve small-angle scattering accounts for only 10% of ρ_0, but doubles $\Delta\rho(\infty)/\rho_0$. This high sensitivity to the details of scattering makes detailed comparison of experiment and theory very chancy, and there seems little point in making the attempt with the data of fig. 28, even though they are the best available. At least one need not be surprised that curve (b), at 4.2 K where phonon effects are small, lies so far above the theoretical for catastrophic scattering. Indeed, the implication is that oxidizing the impurities removes them almost completely from the scene (as witness the great enhancement of r_0) leaving a relatively minor component of small-angle scattering.

Copper serves also to illustrate the effect of small-angle scattering on the transverse magnetoresistance, for which rather little experimental information is available. If catastrophic scattering alone is present, with $\mathbf{B}$ lying exactly along [100] to give hole orbits, as in fig. 14, and electron orbits, the distribution of extra electrons when the Hall angle is large leads to a sharp discontinuity at the boundaries between the two types; the peaks of the electron and hole distributions are diametrically opposite one another. As with longitudinal magnetoresistance, small-angle scattering smooths out the discontinuity, and very little is needed to make a significant difference. It is not so easy, however, to estimate what that difference would be since, as earlier arguments have illustrated, ρ_{xx} is much more sensitively dependent on details than is ρ_{zz}. A model calculation, too long to be reproduced here, suggests that with diffusive scattering alone ρ_{xx} increases without limit as $B^{1/2}$. This behaviour may be expected whenever the Fermi surface results in electron and hole orbits running alongside each other, and may be one cause for the frequent failure of ρ_{xx} to saturate.[36]

This mechanism cannot be immediately invoked to explain the obvious sensitivity to small-angle scattering shown by aluminium (fig. 30). The fact that the sample used was a polycrystalline wire may make detailed interpretation hard but does not invalidate qualitative arguments, since there are no open orbits as in copper, for which the transverse magnetoresistance of a polycrystal is an average of wildly different single crystal curves. Not so in aluminium, however, where all orbits are closed, and electron and hole orbits belong to different bands. The effects of small-angle scattering are to be sought, therefore, either in the individual bands or, as seems a more compelling explanation, as the consequence of interband scattering. Ashcroft's discussion of the band structure of aluminium shows that the Harrison construction is a good starting point, and that the large hole surface in the second zone runs very close to the square rings of electron-like surface in the third zone. The conductivity is dominated by the hole surface, and clearly any scattering from it onto the electron surface may be very disruptive since the next scattering process may bring the

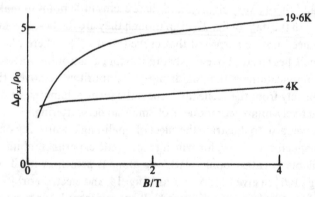

Figure 3.30 Transverse magnetoresistance, at two temperatures, of a wire of polycrystalline aluminium (Fickett[35]); $r_0 = 14.4 \times 10^3$ at 4 K, 3.5 × 10^3 at 19.6 K. The Kohler slope at the high-field end is about 5 times as great at 19.6 K as at 4 K.

particle back to any of three different points on the hole surface. Schematic models have been analysed[36] in which an electron is scattered back and forth across a zone boundary, while remaining on the same replica of the free-electron sphere; while differing in detail they show, as expected, that when **B** is small and the electron spends plenty of time near the zone boundary the orbits get thoroughly scrambled to make each encounter with the boundary the termination of a free path, or something very like it. At high fields the electron rushes past so rapidly that it tends to remain on its proper orbit and the encounter makes less difference. On the other hand, the number of encounters increases in proportion to *B* and the net effect is a field-independent contribution to the scattering. The detailed working-out of the models involves the solution of coupled integral equations which, while presenting no great difficulty, is lengthy enough to be set aside here. The conclusion is that small-angle scattering can result in a considerable enhancement of transverse magnetoresistance in aluminium. Whether this is also the cause of the slow rise at 4 K (see fig. 30), where phonon scattering must be extremely weak, is uncertain since magnetic breakdown, a main topic of the next chapter, occurs in aluminium and is another destroyer of saturation.

174

   The situation in indium is probably similar to that in aluminium. Garland and Bowers[37] have extended the range shown in fig. 1.4 so as to reveal substantial failures of Kohler's rule. With very pure indium the magnetoresistance does not saturate, but rises steadily in high fields, especially when phonon scattering is increased by raising the temperature. And addition of impurity to make catastrophic scattering dominant almost suppresses the high-field rise. This is as clear an example as any of the effects produced by small-angle scattering.

9

   Finally, the influence of small-angle scattering when open orbits are present tends to reduce, rather than increase, the high-field resistivity. This is hardly surprising since the quadratic rise of ρ_{xx} is sharpest when the open orbit conductivity is greatest. Qualitatively the mechanism is the same as operates between electron and hole orbits lying alongside; scattering from the open orbit into a neighbouring closed orbit effectively terminates the free path. Pacher[38] found that heating a copper sample a few degrees from 4 K in a strong field caused an increase of resistance when **B** lay along [100] (closed orbits only) but a decrease when **B** was tilted a few degrees to establish open orbits. This is in accord with expectation, but the information given is too scanty to allow anything more to be said.

110

4 Quantum effects

Hitherto we have managed to avoid taking explicit account of quantum mechanics, even though the passage of electrons through the ionic lattice, and the exclusion principle without which there would be no Fermi surface (and indeed no ionic lattice), are essentially quantal phenomena. But we have made good progress on the assumption that once the dynamical properties of the electrons are established by $E(\mathbf{k})$, classical mechanics suffices for the detailed analysis. Now, however, we must take note of some explicitly quantal effects, viz:

<table>
<tr><td>147</td><td>(a)</td><td>oscillation with changing B of the resistivity, as a result of oscillation of the density of states. This is the original Shubnikov–de Haas effect[1] which is usually rather weak in metals.</td></tr>
<tr><td>153</td><td>(b)</td><td>the often much stronger oscillations resulting from magnetic breakdown,[2] in which electrons tunnel between different sheets of the Fermi surface.</td></tr>
<tr><td>227</td><td>(c)</td><td>a relatively unimportant process that is not normally exhibited by bulk metals, but may appear in thin films, especially if their topology has been suitably designed. It is the result of phase coherence in the electron waves scattered by different defects, and the disturbance of the phase relationships by B.</td></tr>
</table>

In this chapter only (a) and (b) will be considered; (c) will make an appearance in chapter 6. A certain amount of preliminary theory is needed.

The free electron in a uniform magnetic field

In Lagrangian mechanics, and hence in quantum mechanics, the Lorentz force $e\mathbf{v} \wedge \mathbf{B}$, which is neither conservative nor dissipative, is introduced through the vector potential $\mathbf{A}$, defined so that curl $\mathbf{A} = \mathbf{B}$ and div $\mathbf{A} = 0$.[3] This is not a unique definition but for the moment that is no concern of ours;

one form is the *linear* (or *Landau*) gauge:

$$\mathbf{A} = (0, Bx, 0).$$ (4.1)

In the absence of $\mathbf{B}$, the momentum $\mathbf{p}$ conjugate to the position vector $\mathbf{r}$ is just $m\mathbf{v}$ for a classical particle, but when $\mathbf{B}$ is present it is $m\mathbf{v} + e\mathbf{A}$, so that the kinetic energy is $(\mathbf{p} - e\mathbf{A})\cdot(\mathbf{p} - e\mathbf{A})/2m$. To translate into quantum mechanics $\mathbf{p}$ is replaced by $-i\hbar\nabla$, so that Schrödinger's equation takes the form, for a free electon ($V = 0$):

$$(\hbar^2/2m)\nabla^2\psi - (ie\hbar/m)\mathbf{A}\cdot\nabla\psi - e^2 A^2\psi/2m + E\psi = 0.$$ (4.2)

If we substitute (1) for $\mathbf{A}$ we find the equation is separable, and ψ may be written as $\Xi(x)\exp i(k_y y + k_z z)$, where

$$(\hbar^2/2m)\Xi'' - [(\hbar k_y - eBx)^2/2m - (E - \hbar^2 k_z^2/2m)]\Xi = 0.$$ (4.3)

This is the equation for a harmonic oscillator centred on x_0, where

$$x_0 = \hbar k_y/eB = k_y/\alpha,$$ (4.4)

and having a force-constant $e^2 B^2/m$, so that the natural frequency $\omega = eB/m = \omega_c$, according to (1.2). The quantized energy levels therefore are:

$$E_n = \hbar^2 k_z^2/2m + (n + \tfrac{1}{2})\hbar\omega_c.$$ (4.5)

In a macroscopic sample each of these levels, for a given k_z, is highly degenerate. Consider a very long strip for which $0 < x < X_0$, and let periodic boundary conditions apply to the y- and z-components, so that the permitted values of k_y and k_z are such that $k_y Y_0$ and $k_z Z_0$ are integral multiples of 2π. Then for the wave-function to be contained in the sample k_y, according to (4), must lie between 0 and αX_0; there are $\alpha X_0 Y_0/2\pi$ allowed values for each choice of n. If, in addition, we constrain k_z to a range δk_z, there are $Z_0 \delta k_z/2\pi$ permitted values, so that the total number of states having k_z thus specified, and a given value of n, is

$$\delta D = \alpha V \delta k_z/2\pi^2,$$ (4.6)

in which V is the volume of the sample, $X_0 Y_0 Z_0$; a factor 2 has been supplied on account of the electron spin.

This result acquires a special significance when the permitted levels (5) are represented in k-space as if the motion in the xy-plane were classical, with k standing for $mv/\hbar$. A classical particle with kinetic energy $(n + \tfrac{1}{2})\hbar\omega_c$ would describe an orbit of radius $[(2n + 1)\alpha]^{1/2}$ and area $(2n + 1)\pi\alpha$. The phase space is divided into annular regions, each of area $2\pi\alpha$ and thickness δk_z for the chosen range of k_z, and each such region contains δD states. The mean density of states in k-space is therefore $V/4\pi^3$, independent of B and the same as when $B = 0$. The magnetic field causes what was in effect a continuous and uniform distribution of states to condense into the quantized orbits, but without altering the mean density. Consequently, the energy of the gas is not greatly perturbed, though there is a residual effect which is responsible for the weak, but important, magnetic properties of the gas.

Let us now apply a steady, uniform electric field $\mathcal{E}_x$; then (2) must be supplemented by a potential energy term $ex\mathcal{E}_x\psi$ on the left-hand side, and (3) by $ex\mathcal{E}_x\Xi$. Contrary to what one might expect, a time-independent solution is still possible, for the new term merely shifts the centre of the oscillator from x_0 to $x_0 + m\mathcal{E}_x/eB^2$ while leaving k_y and k_z unchanged. Alternatively, one may consider that one of the original wave-functions, that happened to be centred on $x_0 + m\mathcal{E}_x/eB^2$, originally had a different k_y, so that $\mathcal{E}_x$ has caused k_y to be changed by $\Delta k_y = \alpha\Delta x_0 = m\mathcal{E}_x/\hbar B$, leaving the wave-function otherwise unchanged. Such a Δk_y is equivalent to a change of electron velocity $\Delta v_y = \hbar\Delta k_y/m = \mathcal{E}_x/B$, which is just the drift that crossed $\mathcal{E}$ and $\mathbf{B}$ produce in a classical charged particle. There is consequently no reason to expect that the Hall conductivity will be sensibly affected by quantization. As to σ_{xx} etc. we can say nothing without considering collisions, but at least nothing very dramatic seems likely.

Orbit quantization

As soon as a periodic lattice potential is added to (2) the mathematical problem assumes an entirely different aspect. The more it is studied the more complicated it becomes, though not in a way that strongly influences the observable phenomena. Some of the complexities will come to light later in the chapter, but fortunately one can deal with the most important issues by rough-and-ready assumptions founded in physical intuition. Thus a lattice potential $V(x, y, z)$ which, in general, is not separable into $V_1(x, y) + V_2(z)$, does not permit the z-coordinate to be expressed solely as $\exp ik_z z$; there is a z-component of force, fluctuating on the scale of the ionic lattice, that prevents this simplification. On the other hand, Bloch's theorem assures us that when $\mathbf{B} = 0$ the mean motion of a particle, considered as a wave-packet, follows a straight line as if the mean value of the fluctuating force can be assumed to vanish. It can be shown,[4] though in no easy or short way, that the assumption made in deriving, for example, (1.22) is a very good approximation. Provided $\mathbf{B}$ is not so strong as to reduce the orbit to something approaching atomic dimensions, the lattice may be replaced by a structureless background against which the electrons move with modified dynamical properties, as expressed by $E(\mathbf{k})$; moreover (and this cannot be taken for granted) the equation of motion can be quantized in the momentum representation, and with **B** incorporated by treating $\hbar\mathbf{k} - e\mathbf{A}$ as equivalent to $m\mathbf{v}$. All this we take for granted, as well as the corollary that k_z may now be treated as a good quantum number, conserved during the free motion of the particle in **B**.

The semi-classical quantization of a closed orbit, by Onsager[5] and independently by Lifshitz, was the prelude to a period of great activity in which the de Haas–van Alphen and related effects were applied to the elucidation of electron motion in metals. A full account of the history and the theoretical arguments has been given by Shoenberg[6] and need not be recapitulated here. Onsager showed that the phase integral $\oint p\,dq$ round an orbit was proportional to its k-space area $\mathscr{A}_k$, so that quantization of the integral according to the rules of Bohr–Sommerfeld–Wilson gave immediately for the allowed orbits:

$$\mathscr{A}_k = (n + \gamma)\,2\pi\alpha, \tag{4.7}$$

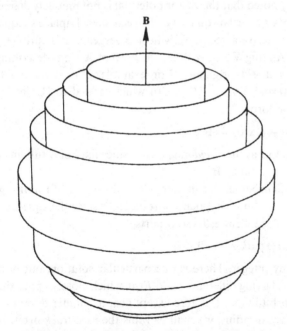

Figure 4.1 Semi-classical representation of allowed states in k-space for a free-electron metal in a magnetic field. Each shell defines values of **k** accessible to electrons in orbits enclosing a given half-integral number of flux quanta, $2\pi\hbar/e$. Only states with energy less than E_F are shown; E_F is the energy of an electron, when $B = 0$, where **k** lies on the circumscribing sphere.

in which γ is a fractional correction, known to be $\frac{1}{2}$ for free-electron orbits, as in (5), but otherwise not deducible except by a searching quantum-mechanical treatment such as that of Roth.[7] This formula applies to the orbits in any plane section of constant k_z, and it implies in three dimensions an array of shells such as that shown in fig. 1 for free electrons. In general the shape of the cross-section of a shell varies with k_z, but the area is constant because of (7), except in a minor way because of fluctuations in γ.

It is somewhat surprising to find that the quantization rule (7) depends on the area of the orbit rather than the perimeter, as one might suppose if the wave-mechanical origin of quantization is the necessity of having an integral number of wavelengths in a complete circuit. In fact the latter requirement is not incompatible with (7), but to appreciate this one must go more deeply into the peculiarities of quantum mechanics.

Gauge transformations

We have already noted that the vector potential is not uniquely defined. In particular from any scalar quantity that satisfies Laplace's equation, $\nabla^2\phi = 0$, we may derive a vector $\nabla\phi$ whose divergence and curl are automatically zero. Adding $\nabla\phi$ to $\mathbf{A}$ does not prevent $\mathbf{A}$ from describing the same field $\mathbf{B}$ as it did originally. For example, if ϕ is chosen to be $-\frac{1}{2}Bxy$, $\nabla^2\phi = 0$ and $\nabla\phi = -\frac{1}{2}B(y, x, 0)$ which, added to (1), leads to a useful alternative form:

$$\mathbf{A} = \tfrac{1}{2}B(-y, x, 0) = \tfrac{1}{2}\mathbf{B} \wedge \mathbf{r}. \tag{4.8}$$

We shall refer to this as the *circulating gauge*, since lines of $\mathbf{A}$ run on circles in the xy-plane, normal to $\mathbf{B}$.

If one sets up Schrödinger's equation, along the lines of (2), in this gauge it is natural to use cylindrical polar coordinates,[8] since the equation then separates into independent z, θ and ρ parts,

$$\psi = F(\rho)\exp\mathrm{i}(k_z z + m\theta), \tag{4.9}$$

in which m is any integer. There is one particular solution, out of many degenerate forms having different m and $F(\rho)$, which closely approximates to the classical orbit in that ψ is zero except near a cylinder of radius $R = |mv/eB|$, the classical radius; we shall call this the *race-track* orbit. It will be appreciated that the degeneracy of the energy levels expressed by (6) means that an unlimited variety of different functions can be constructed by linear superposition, all having equal validity. According to the problem under consideration one may choose the race-track, or the Landau solutions of (3) or, as we shall see, whichever other form may be most convenient. For the present we stick to the race-track and see how it is affected by changes in the gauge of **A**. A brief summary of the relevant points will suffice:

1. If Schrödinger's equation is to hold for all choices of **A**, the addition of $\nabla \phi$ to **A** must be accompanied by a phase change to ψ; $\psi \psi^*$ must remain invariant, and ψ must be replaced by $\psi e^{i\varepsilon}$, where $\varepsilon = \varepsilon(r)$, such that

$$(i\hbar\nabla + e\mathbf{A} + e\nabla\phi)\psi e^{i\varepsilon} = e^{i\varepsilon}(i\hbar\nabla + e\mathbf{A})\psi. \qquad (4.10)$$

It may be verified that ε must be $e\phi/\hbar$ for (10) to hold, and hence for Schrödinger's equation to yield the same solutions apart from un-observable phase changes to ψ. Another way of looking at this point is to write a plane wave $\psi = e^{i\mathbf{k}\cdot\mathbf{r}}$ as $e^{i\varphi}$, and note that $\nabla\varphi = \mathbf{k}$. Now the particle velocity associated with this wave is not $\hbar\mathbf{k}/m$ but $(\hbar\mathbf{k} - e\mathbf{A})/m$, i.e.

$$\mathbf{v} = (\hbar\nabla\varphi - e\mathbf{A})/m.$$

Clearly, to keep **v** unchanged when $\nabla\phi$ is added to **A**, $e\phi/\hbar$ must be added to φ – the same as before; and $e\nabla\phi/\hbar$ must be added to **k**.

2. Let an electron move with velocity **v** round a race-track of radius $R = |mv/eB|$, with the gauge centre at the centre of the orbit, so that on the orbit itself **A** has magnitude $\frac{1}{2}BR$ and circulates in the same sense as the (negative) electron. Then **k** is not $mv/\hbar$ but $mv/\hbar + e\mathbf{A}/\hbar$, i.e. $mv/2\hbar$, or $-\frac{1}{2}\boldsymbol{\alpha} \wedge \mathbf{R}$. The wavefronts are radial and evenly distributed, with wave-length $4\pi\hbar/mv$, round the track. To fit an integral number of waves into the track, $\frac{1}{2}mv^2$ must be $nheB/m$, or $nh\omega_c$. The fractional correction $\gamma, \frac{1}{2}$ for free electrons, represents the extra energy needed to confine the wave-function radially; the radial profile of ψ is just the gaussian function for a harmonic oscillator of natural frequency ω_c in its ground state. The phase length of one complete circuit of the race-track is $\pi\alpha R^2$, or $\alpha\mathscr{A}_r$, where $\mathscr{A}_r$ is the area of the orbit.

3. In addition to the most compact race-track orbit there are other degenerate solutions with different radial variations of ψ and different tangential wave-numbers; and there are also, of course, similar sets of wave-functions differing in energy by multiples of $\hbar\omega_c$. Wave-packets built from these sets can simulate a slightly blurred electron running round the classical orbit. As with wave-packets built from harmonic oscillator functions, these are non-dispersive, holding their original shape inde-finitely, since all components have the same angular velocity, ω_c. Apart from this they are hardly distinguishable from the wave-packets that can be built from plane wave-functions, and we can expect that the scattering by

localized defects will be little affected by the presence of **B**. Such minor differences as are caused by **B** are responsible for the Shubnikov–de Haas effect.

4. The expression for **k** in §(2), being half what one might have expected, raises the question of the condition for Bragg reflection by a periodic lattice. This is usually expressed as a requirement that the component of **k** parallel to the lattice planes be conserved, while the normal component must equal $\frac{1}{2}g$, where $g = 2\pi/a$ and a is the spacing of the planes. Then reversal of the normal component on reflection changes k by $\pm g$, the wave-vector of the reflecting planes. We must now generalize this condition to:

$$\Delta \mathbf{k} = \pm \mathbf{g}, \tag{4.11}$$

$\Delta \mathbf{k}$ being the change of **k** on reflection. It is a consequence of **A** that Bragg reflection of an orbiting electron does not simply reverse the normal component of **k**; this is because the new orbit is centred on a point that is not the centre for the circulation of **A** in (8). We must therefore examine the effect on **k** of moving the gauge centre, to show that the different statements about Bragg reflection are compatible.

5. Suppose originally the gauge centre is at $\mathbf{r}_0$, so that

$$\mathbf{A} = \tfrac{1}{2}\mathbf{B} \wedge (\mathbf{r} - \mathbf{r}_0), \tag{4.12}$$

and is then moved to $\mathbf{r}_0 + \mathbf{S}$. Then **A** is changed by $-\frac{1}{2}\mathbf{B} \wedge \mathbf{S}$, which is constant, so that $\phi = -\frac{1}{2}(\mathbf{B} \wedge \mathbf{S})\cdot\mathbf{r}$ and, from § 1,

$$\varepsilon = e\phi/\hbar = -\tfrac{1}{2}(\boldsymbol{\alpha} \wedge \mathbf{S})\cdot\mathbf{r}. \tag{4.13}$$

Alternatively, as in §1, **k** at any point must be replaced by $\mathbf{k} - \frac{1}{2}\boldsymbol{\alpha} \wedge \mathbf{S}$. This shows that if gauge centre and orbit centre do not coincide the wave-fronts round the race-track are not necessarily radial or evenly spaced.

6. Fig. 2 represents the Bragg reflection at P of an electron from an orbit centred on C_1 to one centred on C_2. If the gauge centre were allowed to

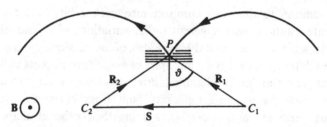

Figure 4.2 Bragg reflection of an almost-free electron from an orbit centred on C_1 to one centred on C_2. The reflecting planes are indicated schematically.

change from C_1 to C_2, the original **k** would be $-\frac{1}{2}\alpha \wedge \mathbf{R}_1$, as in §2, and the final **k** would be $-\frac{1}{2}\alpha \wedge \mathbf{R}_2$. We must not, however, change gauge in mid-calculation, and if we keep C_1 as gauge centre the final **k** will be, from §5, $-\frac{1}{2}\alpha \wedge (\mathbf{R}_2 - \mathbf{S})$ since we have to shift the second gauge centre back through $-\mathbf{S}$. Hence the change of **k** on reflection is $\frac{1}{2}\alpha \wedge (\mathbf{R}_1 - \mathbf{R}_2 + \mathbf{S})$, i.e. $\alpha \wedge \mathbf{S}$. This gives as the condition for Bragg reflection, from (11):

$$\mathbf{g} = \alpha \wedge \mathbf{S}, \tag{4.14}$$

irrespective of the magnitude of R. Now when $\mathbf{B} \to 0$ a free electron with velocity **v** has a de Broglie wavelength $\lambda = 2\pi/\alpha R$ and suffers Bragg reflection if it strikes the reflecting planes at an angle ϑ to them such that $\lambda = (4\pi/g) \sin \vartheta$; ϑ is the angle indicated in the figure, and $\sin \vartheta = S/2R$. Combining these results we find the condition for Bragg reflection to be $g = \alpha S$, as in (14), and compatibility with (11) is demonstrated.

An alternative derivation involves moving the gauge centre to P, the point of Bragg reflection. Application of the rule in §5 shows that the incident wave-vector is $-\alpha \wedge \mathbf{R}_1$ and the reflected wave-vector $-\alpha \wedge \mathbf{R}_2$, the same as for free electrons in the absence of **B**.

7. For the purpose of quantizing an orbit that consists of a number of free-electron arcs joined, as in fig. 2, by Bragg reflections, we need to add up the phase changes suffered by the wave as it traverses the complete orbit. These can be considered as made up of three components: (a) the phase change between successive reflections when gauge and orbit centres coincide, (b) the phase change that must be introduced, as in §6, when the gauge centre is switched by **S** at a Bragg reflection, and (c) the phase change attributable to the precise location of the lattice planes relative to the point of reflection.

To take the last point (c) first, it must be appreciated that if the lattice is moved in the direction of **g**, the point of reflection is unchanged but the phase of the reflected wave goes through 2π for a movement of one lattice spacing. If **g** is the change of **k** on reflection, this point is taken into account by assuming a phase change,

$$\varepsilon_c = (\alpha \wedge \mathbf{S}) \cdot \mathbf{r}, \tag{4.15}$$

where **r** is the position vector of the point of reflection relative to an origin (O in fig. 3) at which the Fourier component **g** of lattice potential has zero phase.

For (b) we need to generalize the result of §6. If the first arc in fig. 3 is gauged with respect to C_1, changing the gauge centre to O introduces a phase change, by the argument of §5, that amounts to $\frac{1}{2}(\alpha \wedge \mathbf{r}_1) \cdot \mathbf{r}$ and the same argument applied to the next arc gives $\frac{1}{2}(\alpha \wedge \mathbf{r}_2) \cdot \mathbf{r}$. It is the difference

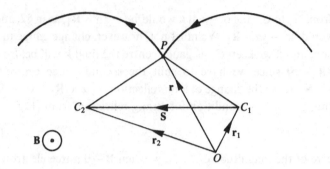

Figure 4.3 Phase change accompanying Bragg reflection at P.

between these two that is the correction (b),

$$\varepsilon_b = \tfrac{1}{2}[\boldsymbol{\alpha} \wedge (\mathbf{r} - \mathbf{r}_2)]\cdot\mathbf{r} = -\tfrac{1}{2}(\boldsymbol{\alpha} \wedge \mathbf{S})\cdot\mathbf{r}, \tag{4.16}$$

so that $\varepsilon_b + \varepsilon_c = \tfrac{1}{2}(\boldsymbol{\alpha} \wedge \mathbf{S})\cdot\mathbf{r}$. This is α times the area of the quadrilateral OC_1PC_2, and is to be taken as positive if a circuit of the quadrilateral in the sense of the letters has the same sense as the electron in its circular orbit.

8.   In a complete orbit made up of arcs, the phase change (a) between two reflections is α times the area of the sector described by the arc, as follows from §2. We can now represent the total phase change round the orbit in terms of areas. Let us note that $\varepsilon_b + \varepsilon_c$ in §7 can be considered as made up of C_1PC_2, which is independent of the choice of O, and OC_1C_2 which is not. In a complete orbit, however, involving n orbit centres $C_1, C_2 \ldots C_n, C_1$, the sum of these origin-dependent contributions is just the area of the polygon $C_1, C_2 \ldots C_n$, and therefore independent of the choice of O.

Two illustrative examples are shown in fig. 4: (a) is an electron orbit, and (b) a hole orbit. That part of (a) starting just before the electron reaches P_1

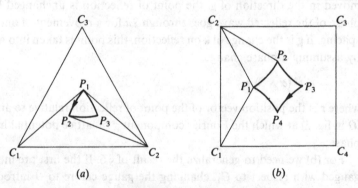

Figure 4.4 Geometrical representation of the phase length of (a) a triangular electron orbit, and (b) a square hole orbit.

and finishing just before it reaches P_2 has C_2 as its orbit centre. Then ε_a is represented by sector $C_2P_1P_2$ (positive), the origin-independent part of $\varepsilon_b + \varepsilon_c$ by the triangle $C_1P_1C_2$ (negative), the two together giving the negative area $C_1P_1P_2C_2$. There are three such negative areas which, added to the polygon $C_1C_2C_3$ (positive), leaves the triangular orbit $P_1P_2P_3$ (positive). The phase change round the orbit is therefore α times the orbit area, agreeing with Onsager's specification (7). The same argument applied to the hole orbit (b), for which polygon $C_1C_2C_3C_4$ is negative, gives a negative area, that of the orbit $P_1P_2P_3P_4$. By these special cases we see that there is no incompatibility between (7) and the notion that orbit perimeter should be the determining factor in quantization; it is just that the latter is incomplete.

The assembly of quantized orbits

After this long, but not irrelevant, digression let us return to (7), the permitted areas of k-orbits. Two alternative ways of looking at this are first, to note that the corresponding orbit areas in real space are:

$$\mathscr{A}_r = \mathscr{A}_k/\alpha^2 = (n + \gamma)h/eB, \qquad (4.17)$$

so that, apart from γ, the orbit contains an integral number of 'flux quanta', h/e; and, secondly, to note that if γ is constant, the energy spacing of successive levels

$$\Delta E \sim (\partial E/\partial \mathscr{A}_k)_{k_z} \Delta \mathscr{A}_k = 2\pi\alpha(\partial E/\partial \mathscr{A}_k)_{k_z} = \hbar\omega_c, \quad \text{from (1.29)},$$
$$(4.18)$$

in accordance with primitive ideas about quantization.

We now consider the effects of quantization on the assembly of conduction electrons. The most widely studied of these is the de Haas–van Alphen effect,[9] oscillations of magnetic moment as **B** changes. Although this only impinges on the magnetoresistance in a few instances, it is worth devoting a short space to the origin of the effect, if only as an illustration of the mode of analysis. It seems at first sight rather paradoxical that the magnetic properties of a metal should be so weak, since every electron is caused to move in an orbit which, by Ampère's theorem, possesses a diamagnetic moment,

$$\mu = - |e\mathscr{A}_r\omega_c/2\pi|, \qquad (4.19)$$

the product of orbit area and circulating current (see (1.32)). Since for an electron of given energy $\mathscr{A}_r \propto B^{-2}$ and $\omega_c \propto B$, μ is inversely proportional to field strength – a remarkable phenomenon, if observable. It is well-known,[10] however, that a classical assembly of electrons has no magnetic properties, since the huge diamagnetic moment of closed orbits is exactly neutralized by the paramagnetic moment of electrons skipping round the periphery of the sample. These 'hole orbits' have so large an area that their quantized levels are unresolvably close, and they contribute a moment of classical magnitude. It is the oscillations about the mean diamagnetic moment due to the quantized closed orbits that are responsible for the observed de Haas–van Alphen effect; we may concentrate on the oscillations henceforward and forget about the cancelling steady components.

   The properties of those electrons whose k_z lies in a narrow range δk_z are shown in fig. 5, in drawing which it is assumed that the quantum number n in (7) is large enough to permit linear interpolation between successive values.* At the top of the diagram (*a*) the energy levels for different n are

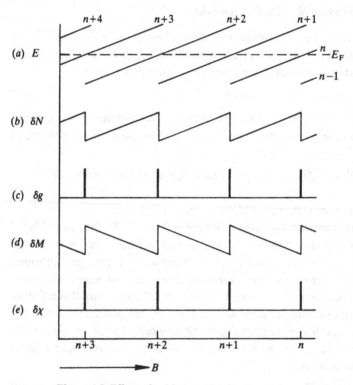

Figure 4.5 Effect of orbit quantization on various physical quantities.

---

* The electron spin, which splits each level into two, is ignored in this treatment. It makes a significant difference to the details of behaviour but is not essential for understanding the physics.[11]

shown rising steadily through the Fermi level, indicated by a broken line and assumed to remain constant (the justification for this assumption will be discussed later). Each crossing is marked on the axis of B as the quantum number of a Fermi electron in this particular slice. In fig. 1 these crossings define the ends of the cylindrical shells. At 0 K every level below E_F is full, the number of states δD, being given by (6), and every level above E_F is empty, so that the number of electrons in the slice oscillates about the mean with the sawtooth form (b). The density of states, $\delta g(E)$, at a chosen energy E is the derivative $(\partial N/\partial E)_{k_z}$ vanishing except when a level coincides with E: the δ-functions in (c) have amplitude δD. If at a given value of B there is a level at $E_F + \Delta, \Delta$ being less than the spacing ω_c,

$$\delta g(E) = \delta D \sum_n \delta(E - E_F - \Delta + n\hbar\omega_c).^* \qquad (4.20)$$

Finally, in (d) the magnetic moment, δM, of all electrons in the slice reflects the variations of δN, $\delta M = \mu \delta N$, μ being given by (19); and the differential susceptibility, $\delta\chi$, in (e) has the same form as δg.

We must now look into the assumption that E_F is constant. Sometimes this is an entirely legitimate assumption, as when the effects of quantization are confined to a small sheet of the Fermi surface, e.g. the square rings in the third zone of aluminium (fig. 3.9), when there is also a substantial reservoir, in this case the second zone hole surface. In copper, however, there is only a single surface and the conservation of particle number is not perfectly compatible with a constant Fermi level. To estimate the magnitude of fluctuations in E_F needed to conserve particle number we must integrate δN over the whole surface, and then see how much E_F changes when the excess or deficiency of N is removed. The mathematical operations involved have a general utility and can be taken over directly into our discussion of the Shubnikov–de Hass effect.

First it is convenient to Fourier-analyse $\delta g(E)$, as given by (20),

$$\delta g = \sum_{p=-\infty}^{\infty} a_p e^{ip\lambda}, \qquad (4.21)$$

where $\lambda = 2\pi (E - E_F)/\hbar\omega_c$ and $a_p = (\delta D/\hbar\omega_c)e^{-2\pi i p\Delta/\hbar\omega_c}$. At temperature T each level is fractionally occupied in accordance with the Fermi function, which may be written

$$f_0 = 1/(e^{\pi\lambda/X} + 1), \qquad (4.22)$$

---

*The reader should be on his guard for a short while not to confuse the Dirac δ-function to the right of $\sum$ with the increment δD to the left.

in which $X = 2\pi^2 k_B T/\hbar\omega_c$. Hence the total number of electrons in the slice

$$\delta N_{tot} = \int f_0 \delta g \, dE = (\hbar\omega_c/2\pi) \sum_p a_p \int^\infty e^{ip\lambda} \, d\lambda/(e^{\pi\lambda/X} + 1). \quad (4.23)$$

The lower limit of the integral is deliberately left unstated; electrons well below E_F occupy totally filled states and play no part in the oscillatory phenomena. A single integration by parts disposes of them as the following section in square brackets indicates, and if we take note only of oscillations about the mean we find for their magnitude

$$\delta N_{tot} = (\delta D/\pi) \sum_{p=1}^\infty (R_T/p) \operatorname{Im} [e^{-2\pi ip\Delta/\hbar_c}], \quad (4.24)$$

in which $R_T = pX/\sinh(pX)$ and describes the reduction of the amplitude of the oscillations because of the width of the Fermi tail.

[Integration by parts of the expression in (23) yields a constant which we ignore and an integral which is dominated by the electrons in the Fermi tail:

$$\int^\infty e^{ip\lambda} \, d\lambda/(e^{\pi\lambda/X} + 1) \to -(i\pi/4pX) \int_{-\infty}^\infty e^{ip\lambda} \operatorname{sech}^2(\pi\lambda/2X) \, d\lambda.$$

$$(4.25)$$

This may be evaluated by integration round a contour that encloses the upper half-plane, in which there are singularities of $\operatorname{sech}^2(\pi\lambda/2X)$ at $\lambda = iX, 3iX$, etc. The poles are of second order, necessitating the expansion of

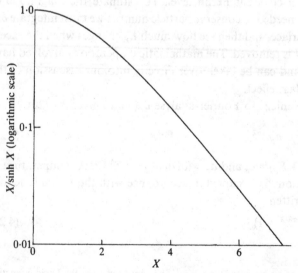

Figure 4.6 The temperature factor, $R_T = X/\sinh X$, for the case $p = 1$.

$e^{ip\lambda}$ in the vicinity of each pole to find the residue. Then the integral in (25) is found to be $(8pX^2/\pi)(e^{-pX} + e^{-3pX} + \cdots)$, i.e. $4pX^2/\pi \sinh(pX)$. On substituting for a_p from (21), (24) is obtained.]

Fig. 6 shows $R_T(X)$ for $p = 1$. The factor $2\pi^2$ in X means that when $k_B T$ is equal to $\hbar\omega_c$ the oscillatory effect has been washed out beyond hope of observation; even when $k_B T$ is only $\hbar\omega_c/4$ the amplitude is reduced tenfold below its value at 0 K. The same factor R_T governs the temperature variation of all the oscillatory effects due to orbit quantization, which are therefore only to be seen at low temperatures. With a cyclotron mass equal to that of a free electron, and in a field of 10 T, $\hbar\omega_c = k_B T$ when $T = 13.4$ K. To see oscillations in UPt$_3$, with a cyclotron mass 90 times greater, the temperature had to be reduced below 0.05 K when B was 11 T. On the other hand, some of the oscillations in bismuth persist well above helium temperatures, and were first discovered at 14 K. The temperature variation of oscillatory amplitude is one of the best methods for determining cyclotron mass, though occasionally it is misleading.

The higher harmonics ($p > 1$) of the oscillations are lost at correspondingly lower temperatures, and it is not unusual to see virtually undistorted sinusoids. There is no need to look further into the waveform here; it is fully discussed in Shoenberg's book.[12]

We proceed to sum (24) over all slices, taking each Fourier coefficient separately, so that $R_T/\pi p$ can be treated as a constant; the rest of the expression leads to a sum of exponential phase factors whose amplitude δD is proportional to δk_z. The phase itself, when n is large, varies rapidly with k_z except near extremes of $\mathscr{A}_k$ where Δ becomes stationary. Measuring k_z from such an extremum, and writing $\mathscr{A}_k''$ for the value of $|\partial^2 \mathscr{A}_k/\partial k_z^2|$ at the extremum, we have

$$\mathscr{A}_k = \mathscr{A}_{k,\text{ext}} \pm \tfrac{1}{2} k_z^2 \mathscr{A}_k'', \tag{4.26}$$

the positive sign applying to a minimum and the negative to a maximum. On a phase–amplitude diagram each slice contributes a vector whose length is proportional to δk_z and whose phase differs from that at the extremum by an amount proportional to k_z^2. When n is large enough the resulting diagram is the Cornu spiral[13] of Fig. 7, which differs from conventional representations only in that the phase at the centre (i.e. the extremum of $\mathscr{A}_k$) is not taken to be zero. As B is changed, the value of Δ at the extremum changes rapidly and the diagram spins about the origin. According to (24) it is the imaginary part of the sum, represented by ZZ', that determines the oscillations of N_{tot}; the length of ZZ' is therefore a measure of the oscillatory amplitude. We ignore the phase except to note

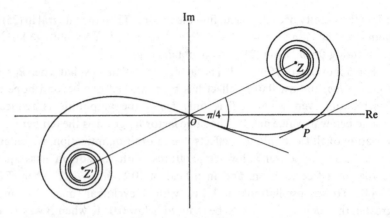

Figure 4.7 Cornu's spiral.

that the resultant lies at a constant angle, $\pi/4$, to the tangent at the origin, and spins at the same rate as the tangent; thus the whole assembly suffers oscillations of N_{tot} whose period is determined by the extremal area, $\mathscr{A}_{k,\text{ext}}$.

The period of the oscillations, ΔB, follows from (7), being that change in B that alters n by unity. Hence

$$\Delta B = 2\pi e B^2/\hbar \mathscr{A}_{k,\text{ext}}. \tag{4.27}$$

If there are several extrema there are several different periods present. All this supposes n to be large enough for the spiral to wind up close to its terminals Z and Z'; when this condition does not hold (as with Fresnel diffraction from a slit rather than a single straight edge) the resultant must be computed directly.

With large n the magnitude is found by remembering a simple property of the Cornu spiral, that OZ is equal in length to the distance round the spiral from O to P, the point where the tangent is parallel to OZ. At P the phase differs by $\pi/4$ from that at O, and the quantum number by $1/8$. From (7), therefore, $|\mathscr{A}_{k,\text{ext}} - \mathscr{A}_{k,P}| = \pi\alpha/4p$ and, from (26), k_z at P is $\pm|\pi\alpha/2p\mathscr{A}_k''|^{1/2}$. If the Fermi surface had constant cross-sectional area within those limits it would from this strip alone generate oscillations matching in amplitude those contributed by the whole real assembly of electrons. Thus we write for the effective strip width, for the pth harmonic,

$$\Delta k_z(p) = |2\pi\alpha/p\mathscr{A}_k''|^{1/2}, \tag{4.28}$$

and the corresponding degeneracy is, from (6),

$$\Delta D = V(\alpha/\pi)^{3/2}/|2p\mathscr{A}_k''|^{1/2}. \tag{4.29}$$

We can now resolve the question that prompted this analysis. The amplitude of the oscillations of N_{tot}, when E_F is fixed, follows by replacing δD in (24) by ΔD:

$$\tilde{N}_{tot} = \sum_{p=1}^{\infty} (2VR_T/\pi)(\alpha/2\pi p)^{3/2}/|\mathscr{A}_k''|^{1/2}. \tag{4.30}$$

To appreciate the meaning of this, it must be divided by the density of states, to see what change in E_F is needed to keep N_{tot} constant. For this purpose, consider a free-electron metal, for which $\mathscr{A}_k'' = 2\pi$ and $g = mk_F V/\pi^2\hbar^2$. Then the oscillations in Fermi level, relative to the level spacing, have amplitude

$$\tilde{E}_F/\hbar\omega_c = R_T/\pi(8p^3n)^{1/2}.$$

Even at $0\,\mathrm{K}$, when $R_T = 1$, and for the fundamental frequency, $p = 1$, the oscillations are very weak; in copper, when $B = 10\,\mathrm{T}$, $n \sim 6000$ and the variations of E_F amount to no more than 10^{-3} times the level spacing – a matter of no consequence. We continue therefore to use the procedure illustrated in fig. 5, assuming E_F to be constant.

The Shubnikov–de Haas effect[1][14]

In the absence of collisions the Hall current set up by crossed $\mathscr{E}$ and $\mathbf{B}$ is unchanged by quantization, and we might expect the effect of collisions in the high-field limit to be similarly unaffected, in the sense that σ_{xx} ought to be $|\sigma_{xy}|/\omega_c\tau$. A number of careful analyses[15] of this point have been made for the free-electron gas without the physical reasoning behind the complexities of the mathematics being successfully explained by most of the authors. There is one case, however, where the result of a full quantal treatment agrees with a much more naive approach, and this is for oscillations of σ_{zz}, the longitudinal magnetoconductivity.[16] We shall start with this before proceeding to the transverse effect where there is an obscure numerical discrepancy between the treatments. It should be said that if there is to be a discrepancy it is likeliest to occur in the transverse effect, where the circular orbit is directly involved. By contrast, so far as σ_{zz} is concerned, scattering in the xy-plane is irrelevant, only changes of v_z being of consequence; and we have seen that the wave-function is separable into a z-component and a transverse component, the former having the character of a plane wave. It is not very surprising, then, that $\mathbf{B}$ has no effect on the structure of the theory. Such effect as it has resides solely in the oscillations of g, the density of states, and the consequent oscillations of τ.

According to Fermi's golden rule the probability of a transition from a given state to one in another neighbourhood is determined by the matrix element of the perturbation responsible, a point defect for example, and the density of states in the neighbourhood to which the transition takes place. The matrix element, we have argued, is substantially unchanged by orbit quantization, but for catastrophic scattering we must take account of the oscillations of g, the overall density of states. We expect τ to vary inversely as g. Some care is needed, however, since the presence of discrete energy levels means that g is far from being a smooth function of E in any one slice δk_z. The model which has been adequate up to now, in which an impulsive electric field generates new particles at the Fermi surface, must be refined to take account of the width of the Fermi tail. In this way we may ensure that in, for instance, elastic collisions of electrons with impurities the density of states available for their reception is the value of g at precisely their initial energy.

The modification is straightforward – instead of the sharp distinction between filled and empty states assumed in fig. 1.14, we allow for partial occupation in accordance with the Fermi function, f_0 in (22). The fraction $f'(\lambda)\,\mathrm{d}\lambda$ of the new electrons in a range $\mathrm{d}\lambda$ is proportional to $-(\mathrm{d}f_0/\mathrm{d}\lambda)\,\mathrm{d}\lambda$ and, when normalised so that $\int f'(\lambda)\,\mathrm{d}\lambda = 1$, is

$$f'(\lambda)\,\mathrm{d}\lambda = (\pi/4X)\operatorname{sech}^2(\pi\lambda/2X)\,\mathrm{d}\lambda. \qquad (4.31)$$

The rate, δv, at which these electrons are scattered elastically into a slice δk_z is

$$\delta v = C\int_{-\infty}^{\infty} \delta g(\lambda)f'(\lambda)\,\mathrm{d}\lambda$$

$$= (\pi C/4X)\sum_{p=-\infty}^{\infty} a_p \int_{-\infty}^{\infty} e^{ip\lambda}\operatorname{sech}^2(\pi\lambda/2X)\,\mathrm{d}\lambda \qquad (4.32)$$

by use of (21). The normalizing constant C is determined by arranging that when (32) is summed over all slices for $p = 0$ it gives the mean scattering rate $\bar{v}(= 1/\tau)$; in fact $C = \bar{v}/g_0, g_0$ being the overall density of states when $B = 0$. Hence for the oscillatory part of v we have

$$\tilde{v}/\bar{v} = (\pi/4g_0X)\sum_{p\neq0} a_p \int_{-\infty}^{\infty} e^{ip\lambda}\operatorname{sech}^2(\pi\lambda/2X)\,\mathrm{d}\lambda, \qquad (4.33)$$

which has the same form as (25). The evaluation of the integral, and the summation over all slices by means of the Cornu spiral, proceed exactly as before to give (if $\tilde{v} \ll \bar{v}$):

$$|\tilde{\sigma}_{zz}/\bar{\sigma}_{zz}| = |\tilde{\rho}_{zz}/\bar{\rho}_{zz}| = |\tilde{v}/\bar{v}| = \sum_{p=1}^{\infty}(2VR_\mathrm{T}/g_0\hbar\omega_c)(\alpha/\pi)^{3/2}/(2p\mathscr{A}_k'')^{1/2}.$$

$$(4.34)$$

The modulus signs indicate that the precise phase of the oscillations has been neglected.

As before, the significance of this result is best appreciated by applying it to a free-electron metal, when it reduces to $R_T/(2np)^{1/2}$. The most one can expect with copper in a field of 10 T, with $R_T = 1$ and $p = 1$, is something under 1% oscillatory amplitude. Where the Fermi surface has portions of small cross-section the local extremal value of n may also be small; but the oscillations arising here are usually swamped by the conductivity of the major portions of the surface. Weak oscillations in magnesium have been studied by Richards,[17] who also gives a careful account of the theory. It is in semi-metals like bismuth,[1][19] however, and in high-mobility doped semiconductors,[18] behaving like semi-metals, that the whole Fermi surface is small enough for the Shubnikov–de Haas effect to be readily seen. Even here, however, as fig. 8 shows, the amplitude in GaSb is no more than a few per cent. At the left, $n = 13$ which might give, at most, an amplitude of 20%; but the longitudinal effect amounts to only 3%.

If we hope that this elementary attack on the problem of $\tilde{\sigma}_{zz}$ can be taken over unchanged to $\tilde{\sigma}_{xx}$, the immediate conclusion is that they should have the same magnitude relative to their mean values, but should be opposite in sign; for in a free-electron metal $\sigma_{zz} = \omega_c \tau \sigma_{yx}$ while $\sigma_{xx} = \sigma_{yx}/\omega_c \tau$. Since,

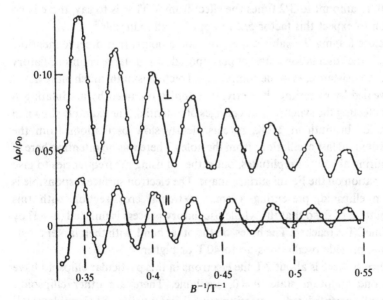

Figure 4.8 Shubnikov–de Haas oscillations in gallium antimonide, for transverse ($\perp$) and longitudinal ($\parallel$) field (Becker and Fan[18]).

however, $\rho_3 = 1/\sigma_3$ and $\rho_1 = \sigma_1/\sigma_2^2$ when $\omega_c \tau \gg 1$, the longitudinal and transverse resistivities should oscillate in phase, as they do in fig. 18. The amplitude of $\tilde{\sigma}_{xx}$ in the diagram is markedly larger than that of $\tilde{\sigma}_{zz}$, in agreement with the searching quantal treatment by Adams and Holstein,[15] who predict $\tilde{\sigma}_{xx}/\bar{\sigma}_{xx} = \frac{5}{2}\tilde{\sigma}_{zz}/\bar{\sigma}_{zz}$. The same sample served for both curves, with current along a [110] direction and **B** either parallel, or perpendicular along [1̄10]. Thus the results are as nearly comparable as can be achieved. It is rather disconcerting, therefore, to find that Sladek,[20] working with a polycrystalline sample of InAs (whose electronic structure is sufficiently near to isotropy to allow useful results to be obtained with a polycrystal) found the amplitude with his strongest fields to be 18% in a transverse field and even more, 24%, in a longitudinal field. It is difficult to comment intelligently on this discrepancy without help from the authors concerning the origin (apart from the involved mathematics) of the factor 5/2. Almost certainly it would be wrong to assume that, while the oscillations of g affect τ for longitudinal conduction to a certain degree, for some unappreciated reason they have 5/2 times the effect on τ for transverse conduction. Bearing in mind that the magnitude of $\tilde{g}/\bar{g}$ and $\tilde{\tau}/\bar{\tau}$ are usually quite small, we should probably look instead for a second process specifically related to the change in the wave-functions, and resulting in a contribution to $\tilde{\tau}$ in the transverse case which happens, for a free-electron metal, to amount to 3/2 times the effect from $\tilde{g}$. That is to say, there is no reason to expect this factor 5/2 to apply to all materials.

Before leaving the subject there are some points that deserve mention. Where the oscillations are superimposed on a large non-oscillatory magnetoresistance, as in the compensated metal bismuth, much more detail is revealed by recording the derivative $d\rho/dB$, obtained by modulating B and detecting the synchronous changes of potential. The example shown in fig. 9, for bismuth at 1.2 K, reveals clearly slow oscillations from the electrons and fast oscillations from the holes. There was no intention here of measuring absolute amplitudes, but rather of using the frequencies to give information on the Fermi surface shape. The electron surface responsible is a thin ellipsoid, presenting a small extremal cross-section with this orientation of **B**, so that the whole oscillatory process is finished ($n \sim 1$) by the time 2 T is reached. The holes, on the other hand, with their larger cross-section, provide oscillations up to 40 T or higher.

Once B exceeds about 2 T the electrons in this particular ellipsoid have only one quantum state, $n = 0$, available. There are other ellipsoids, differently oriented, still possessing several states below the Fermi level,[21] but as B rises further they too reach their *quantum limit*,[22] while the larger

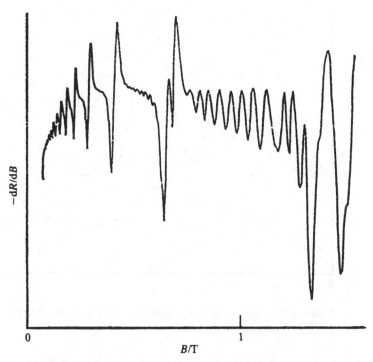

Figure 4.9 Enhancement by derivative technique of the electron (long period) and hole (short period) oscillations in bismuth (Lerner[19]).

hole surfaces in the band below have still some way to go. At this stage, starting perhaps at 40 T, the electron energy will be pushed high enough that electrons are emptied into the lower band and reduce the size of the hole surfaces. Spin splitting of the levels complicates the story but ultimately, in a perfectly compensated sample, all electrons will probably leave the upper band and the material will become an insulator. Only the beginning of the process has been observed,[23] on account of the very high fields required for its completion.

Finally we note the strong oscillations of the longitudinal magnetoresistance when **B** lies along the major axis of one of the electron ellipsoids in bismuth. The surprising aspect of the curves in fig. 10 is that the oscillations are weakened by lowering the temperature. The explanation is given in fig. 11, which is analogous to fig. 1. The current parallel to **B** is carried by an excess of electrons at one end of each cylinder and a deficit at the other, and can only be destroyed by scattering from one end to the other, or between cylinders. In a pure sample, such as was used here, there is still a significant contribution to the resistance by phonon scattering. The

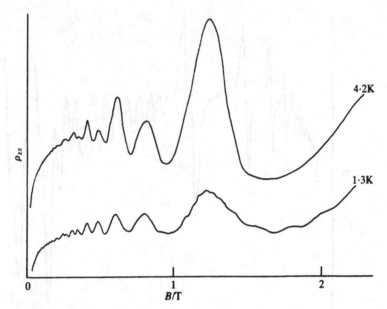

Figure 4.10 Longitudinal magnetoresistance of bismuth at 1.3 K and 4.2 K (Tanuma and Inada[24]).

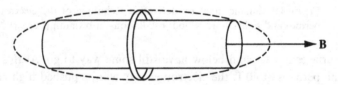

Figure 4.11 Schematic version of fig. 1 near the quantum limit in bismuth. The axial ratio should be considerably greater ($\sim$ 13) than is shown here.

phonon wave-number required to scatter from one end of the long cylinder to the other corresponds to a quantum energy several times k_B/T at 4.2 K, and these processes are rather rare. The important processes are therefore impurity scattering between any available points, and phonon scattering between cylinders (but not end-to-end of the longest). The peaks of ρ_{zz} occur as cylinders are driven out of the ellipse marking the Fermi energy, at which moment (see fig. 5(c)) there is a spike in the density of states and therefore in the scattering rate. The peak at about 1.25 T corresponds to the disappearance of the shorter cylinder in fig. 11. When the temperature is lowered, phonon scattering between cylinders is rapidly reduced and only impurity scattering remains. Since this is rather indifferent as between end-to-end scattering, which shows no spike in g, and inter-cylinder scattering, the oscillations in ρ_{zz} are less marked.

Notwithstanding interesting points like this, the conclusion must be that the Shubnikov–de Haas effect, in metals at any rate, is of rather minor importance. If one seeks strong oscillatory effects in the resistivity one must look to magnetic breakdown as the cause.

Magnetic breakdown[25]

We have already noted that a weak periodic lattice potential which is responsible for Bragg reflection of electrons when **B** is absent may fail to give total reflection when **B** is strong; this is magnetic breakdown. It is convenient to discuss it in terms of almost free electrons that would be executing circular orbits if the weak lattice potential were altogether removed, though Chambers[26] has shown that the same principles apply to the effect of a weak additional potential on an electron otherwise strongly affected by the lattice. Such weak potentials frequently arise from spin–orbit coupling,[27] and as a result the phenomenon of magnetic breakdown is quite widespread.

It is convenient to return to (3) which shows, for the particular gauge chosen, that the x-variation of ψ for a free electron in a magnetic field is the same as if **B** were replaced by a parabolic potential centred on $x_0 = k_y/\alpha$;

$$V_{\text{eff}} = \tfrac{1}{2}m\omega_c^2(x - x_0)^2 = \tfrac{1}{2}m\omega_c^2 x'^2. \tag{4.35}$$

Now suppose we add a weak periodic lattice potential $V'e^{igx}$, which does not disturb the y and z variation but is capable of causing Bragg reflection. The harmonic oscillator wave-function, when n is very large, may be regarded, as in the WKB treatment, as having a spatially varying wave-number $k_x(x)$ such that

$$\hbar^2 k_x^2/2m = E - V_{\text{eff}}, \tag{4.36}$$

and we suppose E to be such that there is a point at which the condition for Bragg reflection, $k_x = \tfrac{1}{2}g$, is satisfied. In the neighbourhood of this point V_{eff} will be assumed to vary linearly with position, so that the kinetic energy $T = E - V_{\text{eff}}$ will also vary linearly. The situation is then precisely analogous to that described as Zener breakdown,[28] in which an electron moves in a lattice under the influence of a uniform electric field, and can tunnel from one band to another through a region where periodic Bloch solutions are not possible. It is this analogy that led to the rather unfortunate term *magnetic breakdown*, but it is much too late to attempt to alter it.

To develop the argument quantitatively, we rewrite (35) and (36) for convenience

$$\hbar^2 k_x^2/2m = \tfrac{1}{2}m\omega_c^2(R^2 - x'^2),$$ (4.37)

where R is the classical orbit radius. The strength of the lattice potential may be defined by the energy gap ΔE which it generates at $k_x = \tfrac{1}{2}g$. From (37) it follows that there is a stretch $\Delta x = \Delta E/m\omega_c^2 x_1'$ through which the electron must tunnel to pass from one band to the next; x_1' is the value of x' at which the Bragg condition is satisfied.

Within this stretch the decay parameter for ψ varies from 0 to a maximum μ_m and back to 0, according to

$$\mu^2/\mu_m^2 = 1 - 4\xi^2/(\Delta x)^2,$$ (4.38)

in which

$$\xi = x' - x_1'$$

and

$$\mu_m = m\Delta E/\hbar^2 g.$$ (4.39)

According to WKB theory the amplitude of ψ is attenuated by a factor $\exp[-\int\mu \mathrm{d}x]$, so that the probability P that an electron will tunnel through is

$$P = \exp\left[-2\int\mu \mathrm{d}x\right] = \exp(-\tfrac{1}{2}\pi\mu_m\Delta x),$$ (4.40)

on substituting (38) and integrating.

At this stage it is helpful to revert to the circular orbit description, using a circulating gauge centred on the point of Bragg reflection so that locally $\mathbf{k}$ takes the same value, $mv/\hbar$, as when $\mathbf{B} = 0$. Then if at this point the electron is moving at an angle θ to the x-axis;

$$k_x = (mv/\hbar)\cos\theta = \tfrac{1}{2}g \quad , \quad \text{and } x_1' = R\sin\theta.$$ (4.41)

Hence,

$$\Delta x = \Delta E/m\omega_c^2 x_1' = \Delta E/eBv_y \quad , \quad \text{where } v_y = |v\sin\theta|,$$

and

$$\mu_m = \Delta E/2\hbar v_x \quad , \quad \text{where } v_x = |v\cos\theta|.$$

From (40), therefore,

$$P = \exp(-B_0/B),$$ (4.42)

where B_0, the breakdown field, is $\pi(\Delta E)^2/4\hbar e v_x v_y$. It is perhaps rather surprising to find that (42) requires no pre-exponential coefficient, but Blount[29] has shown that this is an exact result.

An alternative form of B_0 is in terms of the level separation $\hbar\omega_{c0}$ of free electrons in a field B_0 and the electron energy $E = \frac{1}{2}mv^2$;

$$\hbar\omega_{c0} = e\hbar B_0/m = (\pi/4\sin 2\theta)(\Delta E)^2/E. \tag{4.43}$$

Apart from a factor, usually of order unity, $\hbar\omega_{c0}$, ΔE and E are in geometrical progression. If B_0 is to be, say, 5 T so that breakdown may readily be observed, and $E = 10$ eV, $\hbar\omega_{c0} = 2.9 \times 10^{-4}$ eV, and $\Delta E \sim$ 0.05 eV. This is small by the standard of gaps resulting from the electrostatic lattice potential, but not for spin–orbit effects at points on the zone boundary where otherwise no gap would be expected; examples are found, as we shall see, in Be, Mg, Zn, Al, Sn, Cd and many other metals.

A complete theory of the effects caused by magnetic breakdown presents severe difficulties, even for nearly free electrons, and we shall not attack the problems head-on, but by refinement from a simplified model which nevertheless yields interesting results, even in its crudest form.

The network model[30]

By analogy with the classical picture which has proved so valuable in many applications, we imagine an electron to move in a circular orbit in real space (neglecting the z-dimension) until its direction of motion relative to some set of lattice planes allows Bragg reflection. It then has a probability P of carrying on, and a probability $Q = 1 - P$ of suffering reflection into a new circular orbit. Since there will be at least one other point in the second orbit at which it may once more suffer partial reflection its trajectory lies on a set of coupled orbits, with a statistical distribution among all the possibilities arising from the choice presented at each junction exhibiting magnetic breakdown. When $Q = 1$ and Bragg reflection is complete, we get the situation discussed earlier, an orbit composed of a number of circular arcs, and with no element of choice. We now extend the idea, without at first considering the quantization that is a consequence of phase coherence round the orbit. If there is no coherence it is a reasonable guess that the electron will approximate to a classical particle.

An elementary example is illustrated in figs. 12(a) and (b), in which there is only one set of lattice planes to consider, shown here as running horizontally. The Brillouin zone boundaries in (a) lie at $k_y = \pm\frac{1}{2}g$ and intersect the Fermi surface of the free electron. The periodically replicated pattern in k-space, turned through $\pi/2$ and scaled as usual by the factor $1/\alpha$, gives the r-space pattern (b). When neither P nor Q is zero the electron has a choice of path at each junction, P to proceed, Q to be reflected, and may thus wander along the chain. The resulting change in its effective path must

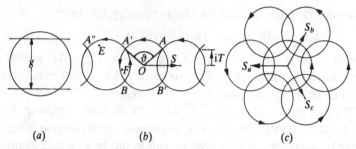

(a)   (b)   (c)

Figure 4.12 Coupling of orbits by magnetic breakdown; (a) a free-electron Fermi surface is intersected by two planes, $k_y = \pm \frac{1}{2}g$, on which weak Bragg reflection occurs; (b) the resulting linear chain of coupled orbits in real space; (c) part of an infinite hexagonal array of coupled orbits.

clearly influence the resistivity. The switching vector **S**, as in (14), has magnitude g/α or $2\pi/\alpha a$ if a is the separation of the lattice planes, and is independent of the energy of the electron.

Fig. 12(c) shows a more complicated, and also more interesting, case appropriate to the hexagonal metals Mg, Zn, etc., where three equivalent sets of lattice planes intersect at $60°$ and generate a two-dimensional set of coupled orbits, with two basic switching vectors $\mathbf{S}_a$ and $\mathbf{S}_b$; a third, $\mathbf{S}_c$, is $-(\mathbf{S}_a + \mathbf{S}_b)$.

The periodicity of the orbit lattice greatly simplifies the calculation of the effective path. Let us suppose $\omega_c \tau$ to be so large that scattering may in the first instance be neglected. Then all electrons starting on a given arc in fig. 12(b) will reach the next junction and thereafter be equivalent. Let us assume the mean terminal of all electrons moving on the arc AA' to be some point E, and of all moving on $A'B$ to be F. Further, let us use complex numbers to denote the coordinates of these points. Then, by symmetry, electrons on BB' have their mean terminal at $-E$, while those on $B'A$ have it at $-F$. If we take S as real and positive, the coordinate of A is $\frac{1}{2}S + iT$, and that of A' is $-\frac{1}{2}S + iT$, etc. Now we note that an electron on AA' has probability P of tunnelling onto $A'B$, when its terminal will be F, and probability Q of being reflected onto $A'A''$, when its terminal will be $E - S$. Consistency requires that the weighted average of these two be E, so that:

$$E = PF + Q(E - S). \tag{4.44}$$

The same argument applied to an electron on $A'B$ shows that:

$$F = -PE - Q(F + S). \tag{4.45}$$

Solving these two equations we find:

$$E = -QS/P \quad \text{and} \quad F = 0. \tag{4.46}$$

If breakdown is complete $(Q = 0)$, $E = 0$ and all electrons have their terminals at the centre of their orbit, as expected when $\omega_c \tau$ is infinite. At the other extreme $(P = 0)E$ is infinite; electrons on AA' are on an open orbit extending indefinitely. It is a little surprising that those on $A'B$ have their terminal at the origin, outside the lenticular orbit that they are confined to. The explanation is that however small P may be, provided it is not zero, there is a chance of the electron leaking out onto the open orbit, where it makes a very large contribution to the mean terminal.

In the presence of catastrophic scattering the calculation is somewhat longer, since allowance must be made for electrons lost along the arcs. But the same procedure works and one of the results is worth quoting: when $\omega_c \tau$ is large,

$$\text{Re}[E] \sim -(QS/P)/(1 + \vartheta/\omega_c \tau P), \tag{4.47}$$

in which ϑ is the angular length of the arc AA'. This is the sort of result one would expect for the combination of two independent scattering processes. Taking $- \text{Re}[E]$ as the effective path L, we have

$$1/L = 1/L_0 + 1/L_1,$$

where $L_0 = QS/P$ as in (46) and represents the behaviour in the absence of scattering, while $L_1 = 2Ql \sin(\tfrac{1}{2}\vartheta)/\vartheta$ and is proportional to l. We shall not have occasion to use this result, but it serves to indicate that when l is much larger than S (i.e. $\omega_c \tau \gg 1$), and there is at least a small breakdown probability $(P \neq 0)$ it will usually be safe to ignore scattering. The dominant random process is now the choice of paths at each junction, and it is this that controls the resistivity.

The incoherent hexagonal network

Magnetic breakdown can have a dramatic effect on the resistivity, as fig. 1.10 shows. When **B** lies along the hexad axis in Zn, the orbits in the central plane are coupled much like those in fig. 12(c), except that the triangular orbits are very small. It is their quantization that generates the slow oscillations. In Mg the triangles are larger and the oscillations seen in fig. 13(a) correspondingly faster; their area is nevertheless only 3.6×10^{-3} times that of the free-electron orbit. Only with exceptionally pure and strain-free crystals $(r_0 > 10^5)$ were the oscillations as large as this; warming to room temperature and recooling reduced them to the size in (b). It was also necessary to lower the temperature well below $4\,\text{K}$, since otherwise the oscillations disappeared, leaving the general run of the curve otherwise unaltered.

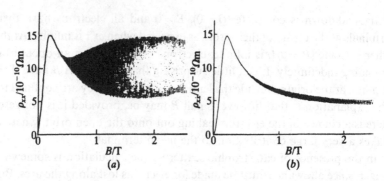

Figure 4.13 Transverse magnetoresistance of a very pure magnesium crystal, with **B** along the hexad axis (Falicov and Stark[25]); (*a*) unstrained, (*b*) after warming to room temperature and recooling.

105   The shape of the smoothed curve is easily understood. Zn and Mg are compensated metals, and the low-field rise approximates to the expected quadratic magnetoresistance. When breakdown occurs, electrons that previously moved in hexagonal hole orbits have a chance of completing circular orbits which are electron-like, and compensation is broken. The precise high-field behaviour must depend on the number of holes that can be converted into electrons, and this can only be known from detailed analysis of the electronic structure. It is possible that B_0 varies with k_z, having a minimum on the central plane, $k_z = 0$; then the thickness of the slice taking part in breakdown will increase with B. On the other hand, and this is the most likely supposition for Mg, B_0 may vary little with k_z, to make the thickness of the breakdown slice independent of B, being determined by the 'necks' of the monster in fig. 3.12 (more correctly, the rounded-off necks in real Mg rather than Harrison's construction). The latter hypothesis is easier to use for computation and turns out to give satisfactory agreement with experiment. It may be noted that in a weak field ($B \ll B_0$) ρ_{xx} depends on sample purity through r_0 (see 1.44), but the high-field resistivity is controlled by the randomness of the breakdown. Only when r_0 is very high does ρ_{xx} show a peak as in the diagrams.

     An additional complication in Zn, compared with Mg, is that, because it is heavier, spin–orbit interaction is much stronger, and it is this that is responsible for the double peaks in fig. 1.10. We shall consider only the case of Mg, Zn having received rather full treatment elsewhere.[31] As with the linear chain we shall first disregard the phase-coherence of electron waves round the triangular orbits, concentrating on the general shape of $\rho_{xx}(B)$. The oscillations will be the subject of the next stage of refinement. We also

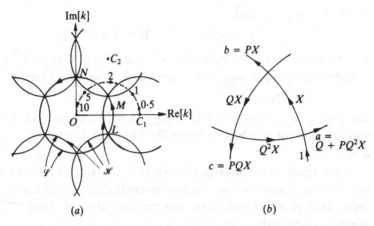

Figure 4.14 Calculation of effective path and conductivity for magnesium, in the absence of phase coherence.

take the triangles to be so small that all electrons may be assumed to start on one of the sides of the hexagonal orbit, $\mathcal{H}$ in fig. 14(*a*). We must not, however, overlook the important role of the triangles, however small, in determining the probability (*a, b* or *c* in fig. 14(*b*)) that an electron entering at the lower right hand will emerge along each of the three possible routes.

To analyse the triangular set of junctions, let there be unit flux of electrons entering and let *a, b* and *c* be the emergent fluxes in the steady state ($a + b + c = 1$). If the flux on the right-hand side of the triangle is X, the other fluxes can be written down immediately, and X found by making the input corner self-consistent, i.e. $X = P + Q^3X$. Hence:

$$a = Q(1 + 2Q)/(1 + Q + Q^2) = 1 - b(1 + Q) \left.\right\}$$
$$b = (1 - Q)/(1 + Q + Q^2) \qquad\qquad\qquad (4.48)$$
$$c = Q(1 - Q)/(1 + Q + Q^2) = Qb. \left.\right\}$$

Much tedious writing is saved by expressing the probabilities in terms of *b*.

Turning now to the network, we follow the procedure used for the linear chain, but this time working in *k*-space. All junctions are identical, and we suppose that the terminal point for electrons leaving L along LM, as shown by the arrow, is designated by the complex number Tk_F. Of those reaching M (all if we ignore collisions) a fraction *c* return to L along the other side of the lens and have their terminal at $k_F(\sqrt{3} - T)$, since $OC_1 = \sqrt{3}k_F$ with this arrangement. A fraction *b* continue along the same circle and have their terminal at $k_F T e^{i\pi/3}$, while a fraction *a* proceed along the circle centred at $C_2 = \sqrt{3}k_F e^{i\pi/3}$, and have their terminal at $k_F(\sqrt{3}e^{i\pi/3} + T e^{-i\pi/3})$. Self-consistency requires that

$$T = c(\sqrt{3} - T) + bTe^{i\pi/3} + a(\sqrt{3}e^{i\pi/3} + Te^{-i\pi/3}).$$

Solving for T and substituting for a, b and c from (48) we find, after considerable manipulation, that T can be expressed very simply,

$$T = \tfrac{1}{2}\sqrt{3}(1 + e^{i\psi}), \tag{4.49}$$

where $\tan\psi = 2\sqrt{3}Q(1 - Q)/(2Q^2 + 2Q - 1)$. The trajectory of T is the semicircle shown as a broken line in fig. 14(*a*), and labelled by values of B/B_0.

In real space electrons starting anywhere with k lying on the arc LM will have a terminal which is $- ik_F T/\alpha$ from the centre of the circular orbit, while those starting on KL will have their terminal at $(- ik_F T/\alpha)e^{-i\pi/3}$, and similarly for the other arcs.

There is no need to enter into the details of calculating the mean starting-point for the electrons excited by $\mathscr{E}$ on each arc, and hence evaluating $\Delta\sigma$ by use of (1.37). The result, in complex notation ($\sigma = \sigma_{xx} + i\sigma_{yx}$) is

$$\Delta\sigma = - i(9\, nek_0/4\pi Bk_F)(T - \tfrac{1}{3}\pi), \tag{4.50}$$

in which $\Delta\sigma$ is the conductivity due to the breakdown zone, of thickness k_0 and n is the number of electrons per unit volume, 2 per atom; $n = k_F^3/3\pi^2$. This must be supplemented by the conductivity contributed by all electrons outside the breakdown zone; because by themselves they are not perfectly compensated and move in closed orbits they add a real part proportional to $1/B^2$ and an imaginary part proportional to $1/B$, as in (3.28). Now when $B \ll B_0$, and $T \sim \sqrt{3}$, the metal is compensated and $\text{Im}(\sigma) = 0$. This gives us the coefficient of the imaginary term in $1/B$, but we leave that of the real term undetermined, writing for the whole metal:

$$\sigma = C/B^2 - i(9nek_0/4\pi Bk_F)(T - \sqrt{3}). \tag{4.51}$$

The value of C is governed by the purity of the sample and must be adjusted to fit the low field behaviour of ρ_{xx}, which is $\text{Re}[1/\sigma]$. The other adjustable constant, k_0, can be estimated from the high-field behaviour, when the term C/B^2 is negligible for the samples used to obtain fig. 13. Computation from the high-field end of the curve indicates that ρ_{xx} is tending to $3.9 \times 10^{-10}\,\Omega\text{m}$, but convergence to the limit is low. When $B \gg B_0$, $Q \sim B_0/B$ from (42), and $b \sim 1 - 2B_0/B$ from (48). Substituting in (49) and (51) we find:

$$\rho_{xx}(\infty) = 4\pi k_F B_0/9nek_0. \tag{4.52}$$

This leads to $k_0/k_F B_0 = 0.27\,\text{T}^{-1}$; if the estimate of $0.585\,\text{T}$ for B_0 (which is none too reliable) is assumed, k_0/k_F becomes 0.16, and finally (51)

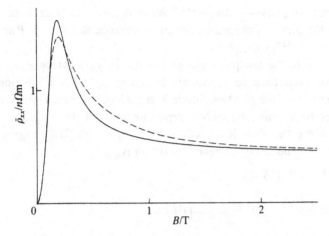

Figure 4.15 Transverse magnetoresistance (full line) calculated for the model of fig. 14, compared with experimental curve, fig. 13(b) (broken line).

may be written, to apply to fig. 13,

$$\sigma = 1.2 \times 10^7/B^2 - \mathrm{i}(2.5 \times 10^9 B_0/B)(T - \sqrt{3})\,\Omega^{-1}\,\mathrm{m}^{-1}. \qquad (4.53)$$

The value of k_0/k_F compares quite well with a direct measurement,[32] by oscillations of ultrasonic attentuation, which gave 0.176 instead of 0.16. It is not out of the question that B_0 is 0.65 T rather than 0.585 T, in which case there would be perfect agreement for the waist dimension. The curve for ρ_{xx} derived from (53) is shown in fig. 15, together with the smoothed experimental curve. Apart from the width of the peak agreement is rather good, and we may proceed in the confidence that the general concept is satisfactory. Probably the assumption that the same k_0 applies to all values of B is over-simple, but no attempt will be made to invent a better model.

Local coherence on the hexagonal network

The frequency of the oscillations in fig. 13 shows that they arise from quantization of the small triangular orbits, and that therefore the electron waves maintain coherence round the triangles. It is easy to take account of this by reconsidering the system of junctions in fig. 14(b). When a wave of unit amplitude arrives at a junction an amplitude p continues along the same free-electron orbit, while there is an amplitude q of Bragg-reflected wave. Particle conservation demands that

$$pp^* + qq^* = 1, \qquad (4.54)$$

and p and q are in phase quadrature.[30] We shall take q to be real and p imaginary – the details of phase are of small importance. Since $pp^* = P$ and $qq^* = Q$, $|p| = e^{-B_0/2B}$, from (42).

Turning now to the complete triangular orbit in fig. 14(b) we allow a wave of unit amplitude to arrive at the lower right-hand junction, generating a wave of (complex) amplitude X immediately after the junction. What arrives back after one circuit, replacing $Q^2 X$ in the diagram, is $q^2 X\, e^{i\phi}$, where ϕ is the phase length of the orbit, $\mathscr{A}_k/\alpha$ from (7) if we ignore the correction γ. The self-consistency condition is:

$$X = p + q^3 X\, e^{i\phi},$$

so that

$$XX^* = P/(1 - 2Q^{3/2} \cos\phi + Q^3)$$

and

$$b = P^2/(1 - 2Q^{3/2} \cos\phi + Q^3). \tag{4.55}$$

The transmission coefficients a and c are related to b exactly as in (48). When B/B_0 is small and Q nearly unity the denominator in (55) is a typical resonant denominator that produces narrow peaks, when $\phi = 0$, in b and c, of magnitude 4/9, while a drops to 1/9 in spite of almost perfect Bragg reflection. At higher field strengths the oscillations of the coefficients become more nearly sinusoidal. The mean values of these coefficients agree with those in (48).

In practice the oscillation amplitude is controlled by several factors. The temperature factor, R_T in (24), reduces the higher harmonics and may well reduce the fundamental considerably at the low-field end; also it is not out of the question that strains and other inhomogeneities may smear out the oscillations, as discussed fully by Shoenberg.[33] The third point, however, is more fundamental and applies to the breakdown oscillations as much as to the Shubnikov–de Haas oscillations already discussed – the phase depends on the precise cross-section of the triangular orbit, which varies with k_z. One must therefore know the Fermi surface shape in detail in order to apply the Cornu spiral argument that led to (27). Without reliable information we may take the effective thickness of the breakdown slice for the fundamental frequency, $\Delta k_z(p = 1)$, as an adjustable parameter.

Computation of ρ_{xx} follows in a manner similar to the derivation of (53) and thence fig. 15, but if one assumes that only the fundamental frequency matters the line-shape implied by (55) must be resolved into Fourier components, and the whole process is easier to compute than to analyse. De Haas–van Alphen measurements[34] give an estimate of the cyclotron mass of the triangular orbits as one-tenth the free-electron mass, from which R_T is calculated. In order to fit the oscillation amplitude at 2.5 T in fig. 13(*b*) it was necessary to take Δk_z as 0.0164 $B^{1/2}k_F$, having the same field-variation as (28). The resulting variation of amplitude with field then agreed quite well with the experimental curve, except that the latter lost its oscillations below 0.5 T, while theoretically they should have been visible down to 0.35 T. It is worth commenting that the assumed magnitude of Δk_z differs from that predicted by the Harrison model, 0.0146 $B^{1/2}k_F$, by only 12%. There is little point in worrying about such small discrepancies, especially as the status of the experimental results is uncertain. Stark, who pioneered the production of extremely pure crystals of magnesium, has remained, with his students, sole proprietor of a process which has been only sketchily described in print;[35] he has unfortunately released only a tiny fraction of what must be an enormous body of information, especially on the breakdown oscillations just discussed.

The fully coherent network[30]

We now return to fig. 13(*a*), measured on a sample that had been treated carefully to minimize strains. The very large oscillations cannot be explained in terms of the partially coherent network, especially as they sit upon the curve for ρ_{xx} in fig. 15 rather than spanning it more or less evenly, a point noted by Falicov and Stark[25]. They also publish a detailed trace of a few cycles showing clearly a high-frequency component due to the lens-shaped orbit, $\mathscr{L}$ in fig. 14(*a*). Partial coherence at least must now spread through the network to prevent the triangular orbits from being considered as independent units; at this point real difficulties present themselves. No adequate theory of the conductivity of a coherent network has been devised, though the magnetic properties evidenced by the de Haas–van Alphen effect are not so difficult. Even so, they are complicated enough to indicate where the obstacles lie in the path of a complete conductivity theory.

As a preliminary let us note that a perfectly regular and phase-coherent orbit lattice has the same property as an atomic lattice in respect to generating Bloch-like solutions to the wave equation, so that we might expect quasi-particles to move in straight lines despite the presence of **B**. We shall soon discover that the periodicity of the orbit lattice is not the simple matter that it is with an atomic lattice, and that the concept of quasi-particles is open to doubt. Nevertheless it is an enticing prospect, for it takes only a small number of electrons moving in straight lines, with long free paths, to enhance σ_{xx} considerably. So long as $|\sigma_{xy}| > \sigma_{xx}$, i.e. $\omega_c\tau > 1$, as is certainly the case with Mg under the conditions of fig. 13, ρ_{xx} varies in the same sense as σ_{xx}, so that the higher values of ρ_{xx} in (a), compared with (b), might find an explanation along these lines. It is not just the slice Δk_z, responsible for the oscillations, that might enhance σ_{xx}, but all electrons in the breakdown zone. With the parameters we have chosen to fit curve (b) ($\Delta k_z = 0.0164\,B^{1/2}k_F$ and $k_0/k_F = 0.16$) $\Delta k_z/k_0 \sim B^{1/2}/10$ so that at $B = 2.5\,\mathrm{T}$ there are 6 times as many electrons available for enhancement as for oscillation. In the light of this the coincidence of base lines in (a) and (b) would have to be regarded as accidental. This is enough to raise doubts about the explanation and/or the wisdom of reading too much into two imperfectly substantiated observations. We therefore leave explanations aside and concentrate on the general problem of quantizing a perfectly coherent network.

By analogy with the Bloch function in a periodic lattice, e.g. (3.25), we seek a complex function defining the amplitude and phase of the wave everywhere on the network and having the same form in each cell, apart from a phase factor $e^{i\mathbf{k}\cdot\mathbf{r}}$. For the hexagonal network of fig. 16 we might choose the rhombic unit cell shown by heavy lines, but what gauge centre shall we choose? If we work throughout with the same centre, O for example, the cells cease to be equivalent, being differently located relative to O. One might hope to restore equivalence by gauging each arc with respect to its own orbit centre, as discussed at the beginning of the chapter, but this too is seen to fail. For, as fig. 3 indicates, the phase correction at P is α times the area of the quadrilateral OC_1PC_2, and at P' it is α times $OC_1'P'C_2'$, the two areas differing by half the area, Σ_0, of the unit cell of the

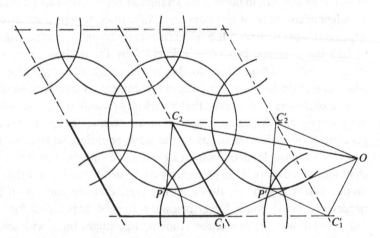

Figure 4.16 Orbit network for the hexagonal lattice (cf. fig. 12(c)) showing, in heavier lines, one choice of unit cell.

orbit lattice. If, however, B is chosen so that $\frac{1}{2}\alpha\Sigma_0 = 2n\pi$, n being an integer, every junction differs in phase correction by a multiple of 2π from its equivalents in other cells, and the orbit lattice is strictly periodic.

It is helpful to note that the orbit lattice is similar to the atomic lattice, as follows by comparing (14) with (3.26). The latter may be applied to a two-dimensional lattice by using α as a dummy vector, normal to the plane. Then,

$$\mathbf{g}_a = 2\pi\alpha \wedge \mathbf{b}/\alpha\Sigma = \alpha \wedge \mathbf{S}_b, \qquad (4.56)$$

in which Σ is $|\mathbf{a} \wedge \mathbf{b}|$, the area of the unit cell of the atomic lattice, and $\mathbf{S}_b$ is a basis vector of the orbit lattice; $\mathbf{a}$ and $\mathbf{S}_a$ are similarly related, from which it follows that

$$\mathbf{S}_a = s\mathbf{a} \quad \text{and} \quad \mathbf{S}_b = s\mathbf{b}, \quad \text{where } s = 2\pi/\alpha\Sigma, \qquad (4.57)$$

thus proving the similarity. The magnetic flux Φ threading unit cell of the orbit lattice is $B\Sigma_0$, i.e. $(h/e)\alpha\Sigma_0$, which is $2n(h/e)$ when the above condition for strict periodicity is satisfied; this is an even number of the 'flux quanta', h/e, which are ubiquitous in quantum situations involving magnetic fields. Since $\Sigma_0 = s^2\Sigma$, $s = \alpha\Sigma_0/2\pi = 2n$ in these circumstances. All equivalent junctions are identically situated with respect to the atomic lattice and therefore introduce the same phase changes at Bragg reflection, while the quantization of flux simultaneously takes care of the periodicity as affected by the vector potential.

Formal treatments[36] by group theory lead to the conclusion that the criterion just derived is unnecessarily restrictive in that any integral number, not just an even number, of flux quanta confers strict periodicity. It is worth mentioning that the arbitrary nature of $\mathbf{A}$ implies that the periodicity is not automatically manifest; any periodicity that may be found can be destroyed by adding the gradient of an aperiodic potential, ϕ. Conversely one has no guarantee that one has discovered the best gauge to reveal the fundamental period. Since the result of group theory can almost certainly be realized by taking enough care with the choice of gauge, and since in any case it is far more important to recognize that B must be restricted than to know precisely what the restriction is, we shall take the matter no further. From now on the assumption will be that strictly periodic solutions, with all cells of the orbit lattice equivalent, are possible only if s is integral and the cell contains an integral number of flux quanta, h/e. Alternatively expressed, the unit cell of the atomic lattice, being smaller in area by s^2, must hold an integral submultiple of a flux quantum. Before proceeding further, let us note that this result is equally valid for a three-dimensional lattice, provided $\mathbf{B}$ lies precisely along the third lattice vector $\mathbf{c}$.

The smallest deviation destroys periodicity. The fact that experimenters have not found it necessary to achieve literally perfect alignment, to get reproducible results, indicates that we must not make too much of the mathematical peculiarity of the problem. Nevertheless it cannot be dismissed as a theoretical artefact, since there appear to be observable consequences, if only in rather special circumstances.

Returning to the two-dimensional case we observe that when B is chosen so that s is non-integral but rational, $s = N/M$, say, where N and M have no common factor, a unit cell of the orbit lattice may be defined to have linear dimensions N times the atomic lattice and contain M^2 of the original orbit lattice cells.* For example if the hexagonal network is being analysed the unit cell must contain M^2 orbits centres and the spectrum of stationary states will be correspondingly more complex, since the repeating unit has a structure consisting of M^2 non-equivalent basic orbits. As B is continuously varied, almost all the time s is irrational and there is no unit cell, but at infinitesimal intervals it passes through rational values, usually with M very large but occasionally with M small, even unity when s is integral. This is an example, uncommon in physics but not of great rarity, where a phenomenon is characterized by two periodicities whose relative magnitude is continuously variable, and the discontinuous changes in the repeat distance, at an infinitely rapid rate, are mathematically inevitable even though they do not advertise their presence in any obvious manner.

The larger the repeat distance of a periodic structure, the more the gaps in its spectrum of energy levels, as is clearly illustrated by the simple case of an electron moving along a line on which there is a weak periodic potential. The parabolic $E–k$ curve is interrupted by energy gaps whenever $k = n\pi/a$, a being the period, and consequently in a given energy range the number of gaps is proportional to a. This example serves to explain the extraordinary diagrams devised by Hofstadter,[38] and later investigated in detail by Wilkinson,[39] purely as a mathematical phenomenon. The diagram in fig. 17 shows, as vertical lines, the bands of allowed energies for which there are travelling wave solutions to the orbit lattice problem, and the horizontal axis is a measure of $s = N/M$ over the interval between successive integral values (in this mathematical model the integral values of s give a continuous spectrum). The values of M are here limited to integers less than 40, but already there is enough to show the typical fractal structure in which small sections mimic in miniature the pattern of larger sections. If

---

* By abandoning the obvious symmetry of the orbit lattice it is possible to reduce the unit cell to one containing only M, not M^2, original cells[37].

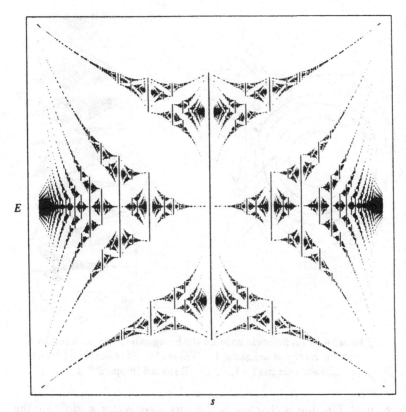

E

s

Figure 4.17 Wilkinson's[39] results for a mathematical model related to the magnetic breakdown model, presented in the form of an energy-band diagram. Vertical lines represent allowed energies, and the horizontal scale of s runs from one integer to the next.

M were unlimited one could magnify any section to any extent and still find the same pattern reproduced down to the smallest detail, apart from the shear distortions obvious in fig. 17. In practice, of course, collisions and imperfections must at some level of detail eliminate the substructure, but if a sample is good enough to show anything that can be ascribed to the presence of this complexity in the spectrum it poses an extremely awkward theoretical problem. We must keep this in mind as we turn to a more intimate examination of the case when s is integral, before reverting to those few experimental results which demand consideration of the non-integral case.

The Bloch functions for a hexagonal network have been evaluated[40] in detail for integral s, to give the typical set of energy bands shown in fig.18.

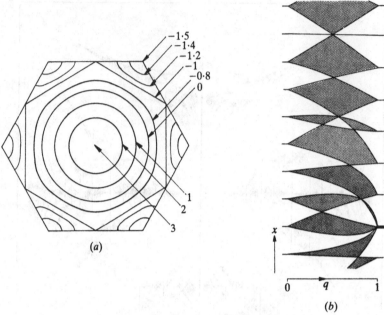

Figure 4.18 (*a*) Brillouin zone for the hexagonal lattice, with contours of constant energy as measured by $C(E)$ in (58); (*b*) typical band structure, with allowed energies $(-1.5 \leqslant C \leqslant 3)$ shaded (Pippard[40]).

Each Bloch function is characterized by its wave-vector κ, defining the phase variation from one cell of the orbit network to another. The Brillouin zone in κ-space is shown in fig. 18(*e*); it has the same form and the same contours as in the tight-binding model of electrons in a hexagonal atomic lattice – if S_a, S_b and S_c are equal vectors at $2\pi/3$ to each other, joining the centres of neighbouring free-electron orbits, a typical contour is defined by

$$\cos \kappa \cdot S_a + \cos \kappa \cdot S_b + \cos \kappa \cdot S_c = C(E), \qquad (4.58)$$

in which C is a function of energy, to be determined by solving the lattice problem. The range of C, for real solutions, is from 3 at the zone centre to -1.5 at each corner. There is no need to quote here the rather complicated functional form of $C(E)$; fig. 18(*b*) suffices to show its essential character for a network that closely resembles what is expected for the central section of the Fermi surface in magnesium. The high-field extreme $(p \gg q)$ is on the left, where the sharp evenly-spaced levels are those of free electrons; on the right, where B is small and $p \ll q$, there are two sets of evenly-spaced levels, the close set for the hexagonal orbits and the wide set for the triangular orbits. The parameter x is a measure of E/B rather than E alone, so that as B is changed to shift from one integral value of s to another the pattern remains substantially the same; but in applying it to a given metal one must remember that the Fermi level appears at a different value of x, and that the appropriate value of q also changes with B. And, of course, as B is changed continuously this diagram, complex as it is, is replaced at extreme rapidity by a sequence of even more complex diagrams, only reverting to this form at every integral value of s.

In measurements of magnetoresistance and the de Haas–van Alphen effect the only part of the energy-level diagram that matters is that around the virtually fixed Fermi energy. The significance of this is easier to see by going back to the Harrison model of the Fermi surface. If we were to ignore the details of magnetic breakdown and draw an energy level diagram for the electrons in the central section the free-electron levels would form an evenly-spaced set above a limiting point, where the quantum number $n = 0$. Its energy would coincide with the bottom of the band, at the centre of the Brillouin zone. The levels for the triangular orbits would extend from a much higher limiting point, for they exist only when the electron energy is high enough to begin populating the next higher zone. High above this is the limiting point for the hexagonal hole orbits, which become smaller, not larger, as E is increased and vanish at an energy well above E_F. As B is raised, each set of levels expands like a concertina from its limiting point, the electron levels moving up and the hole levels down. The relative positions of the sharp levels to either side of fig. 18(*b*) thus change continually, and as each integral value of s is reached the character of the level structure around E_F is different every time.

97

This is a rather unimportant matter so far as the de Haas–van Alphen effect is concerned, or any other effect which depends on the density of states (or some similar quantity) averaged over the Fermi tail. For each set of levels due to a given orbit contributes a periodic variation with B of the measured quantity, such that Fourier-analysis separates each from the others. This is not true when the Shoenberg interaction is important, for this introduces a non-linear mixing of the Fourier components. Without this, however, the concertina movement of the levels is reflected in the periodicity of the components, and their precise relative positions only affect the phases of the components. When magnetic breakdown is present there are many more sets (an infinite number, in principle) of evenly-spaced levels, but the irrelevance (except for phase) of their precise positions still applies. Falicov and Stachowiak[41] have shown how each closed loop that can be drawn on the orbit network contributes an oscillation whose frequency is proportional to its area, and whose amplitude is determined by the probability that an electron will successfully negotiate all the nodes in the loop to return to its starting point. There is no need to go into great detail; the point of interest for the present discussion is that the value of s is irrelevant to the question of which frequencies will be present in the de Haas–van Alphen oscillations. In fact no observations of the effect have revealed anything that has demanded that this opinion be revised.

It is otherwise with the magnetoresistance, although the evidence is scanty, only the observations of Hulbert and Young[42] on tin being sufficiently clear-cut to take into account. In a field of 7.8 T (the caption of their fig. 2 is in error on this point), parallel to the c-axis, there is an oscillation in the resistance whose period is only 2.5 mT, half that of the most rapid oscillation attributable to orbits on the Fermi surface. In fact, this period corresponds, through (27), to an area in k-space equal to the Brillouin zone area. This is just the periodicity with which integral values of s recur. It is noteworthy that the amplitude of these *zone oscillations* is much less temperature-dependent that that of the normal orbit oscillations, so that at 4 K, when the factor R_T has eliminated most of the latter, the former survive at a readily observable level. As Hulbert and Young point out, the special values of B that confer integral values on s are independent of the electron energy, so that all electrons in the Fermi tail give rise to phase-coherent zone oscillations; the factor R_T does not apply here. This is very convincing evidence for the importance of the precise value of s in the magnetoresistance, if not in the de Haas–van Alphen effect.

One can see in a general way why this might be so. As B is changed the Fourier components of the density of states, calculated by the prescription of Falicov and Stakowiak,[41] change their amplitudes only slowly but their relative phases quite rapidly; indeed, the components attributable to long excursions on the orbit network may have, individually, very small amplitude but they are numerous and change phase with great rapidity. It is the sum of the infinity of components, combining in different phases, that generates a fluctuating energy level diagram like fig. 17. All this may pass unnoticed in the de Haas–van Alphen effect, but if there is any validity in the suggestion that broadened energy bands can describe wave-packets moving in straight lines, the breadth of the bands is important, for it determines the group velocity of a typical packet. Analogously to (1.22), $\mathbf{v} \propto \nabla_\kappa E$, and one may guess that the mean velocity is roughly proportional to the fraction of the energy range occupied by the bands, being large when M is small and decreasing rapidly as M increases. Of course, collisions destroy phase coherence and smooth out the rapid alternations, but in a good sample one should find sharp spikes of conductivity when $M = 1$ and s

is integral, with lesser spikes when $M = 2, 3$, etc. It is interesting, in this connection, to note that Hulbert and Young found a strong second harmonic in their zone oscillation when they took particular care in aligning the crystal axis with **B**.

No success has been achieved from attempts to proceed beyond this point in developing a quantitative theory of magnetoresistance in the presence of magnetic breakdown. It would be nearer the mark to say that little, if anything, has been published that bears on the problem. On the other hand, a number of authors have attacked the calculation of energy level structure from various angles, and it is in order to give a brief account of this work. Some of it is extremely dense and hard to follow, while other accounts are too sketchy to admit of a detailed critique. My comments, therefore, should be read as personal, rather than as carrying the consensus of informed opinion which in this matter does not exist.

The status of the theory

It should be clear by now that there is a considerable hiatus between the formal theory, as exemplified by Hofstadter[38] and Wilkinson,[39] and the network model used to interpret individual systems. The model has been fairly successful, but it was set up in a tentative optimism and deserves to be related more soundly to basic principles. The most thorough-going attempt has been by Capel[43] who has demonstrated in several long papers that the one-dimensional chain of circular orbits, as in figs. 12(a) and (b), does truly represent the structure of the formal theory. His methods have not been extended to two-dimensional networks, with their additional complications arising from the precise value of s, but there is no reason to expect any new factor to enter that will cast serious doubt on the network.

166

Even so, there is a serious gap between the calculation of the energy level diagram and the magnetoresistance, especially as the latter demands introducing scattering into the theory, confusing still further an already confused situation. The attempt by Gonçalves da Silva and Falicov[44] is admittedly exploratory and consequently hard to evaluate justly. They treat a limited region of the network as a coherent unit embedded in a matrix whose properties are an average of the properties of different units, not unlike the device we shall meet in the next chapter for dealing with inhomogeneous samples. They derive expressions for the effective path of an electron, but there seems to me to be an unresolved conflict between the electron considered as a classical particle moving on the network and the wave description which, in a coherent unit, must not be reduced to paths on

182

individual arms. The effective path, in fact, is essentially a classical concept. I am therefore disinclined to devote space to an account of their conclusions.

By contrast, Eddy and Stark[45] do not seek to extend the network model but rather to cast doubt on its validity, believing that their measurements of the de Haas–van Alphen effect in magnesium cannot be interpreted in the way suggested by Falicov and Stachowiak,[41] even under conditions that rule out significant disturbance by the Shoenberg interaction.[46] Thus they find the amplitudes of the observed Fourier components to vary rapidly with B, and also see components at frequencies that correspond to no orbit on the network. Unfortunately this work is published with insufficient detail to allow independent assessment; for example, only the Fourier-analysed data are given, with no indication of the procedure used to avoid (or at least quantify) artefacts arising from truncation of the data set. We have here an example of the deplorable modern habit of presenting results in so completely digested a form that the reader must either trust the authors' competence or reject the evidence as being little better than hearsay. In this case I elect the latter course in view of the authors' promise, never delivered, to publish their procedures in full at a later date.

Similarly I am not persuaded by their separate assertion that Fourier analysis of the computed energy-level structure of the hexagonal network reveals components other than those predicted by Falicov and Stachowiak. The latter tested their ideas against the analytical treatment of the linear chain, and I have repeated this test without finding any components whose amplitudes disagreed with their predictions. The density of states varies so sharply with energy that it is highly undesirable to rely on the fast Fourier transform, as did Eddy and Stark. This divides the range to be analysed into equal intervals and may seriously distort the weight to be assigned to critical points at the band edges. Here again one needs to see in considerably greater detail what was actually done before one dispenses with a theory that seems to be well founded and consistent with other approaches.

Lastly, mention must be made of Taylor's[47] treatment of magnetic breakdown in the simplest possible case, the lattice potential being in the form of regularly spaced parallel lines on which the potential is a δ-function. The corresponding network model is a linear chain, but Taylor treats the problem by analogy with the Kronig–Penney theory of electron bands in one dimension. He derives a transfer matrix relating the wave-function in one atomic cell to that in the next, and argues from its form about the presence or absence of particular Fourier coefficients in the spectrum of energy levels. Since in this case there is coincidence between the network model of the density of states and the approach of Falicov and Stachowiak, the discrepancies found by Taylor imply a failure of the network model, contrary to Capel's conclusion. Taylor's argument is partly intuitive, lacking Capel's mathematically more rigorous procedures, and eschews estimates of the amplitudes of the rogue components he believes to be present; it is conceivable that they are vanishingly small. Tempted though I am to side unequivocally with Capel, I must admit I cannot follow his treatment sufficiently easily to be able to give it the attention and critical acument it deserves. I leave the matter as I began, strongly inclined towards the network model but not competent to pronounce authoritatively in its favour.

Other examples of magnetic breakdown

Many of the leading principles of magnetic breakdown have been exposed in the preceding treatment of the hexagonal network, and it remains to indicate briefly a number of other examples without subjecting them to close examination.

1. The linear network in tin

When **B** lies along the c-axis, as in the work of Hulbert and Young[42] already discussed, the network in the central plane of tin is not unlike that in

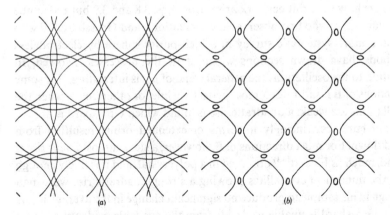

(a)   (b)

Figure 4.19 (*a*) Fermi energy contours in one section of the Harrison construction for tin; (*b*) essential features after introducing small energy gaps (Barklie and Pippard[48]).

magnesium except that it has square rather than hexagonal symmetry (fig. 19). The simplified version (*b*) shows how the principal orbits are coupled by smaller orbits in a way similar to the coupling of hexagons by small triangles in fig. 12(*c*). Only fairly near the central plane is the breakdown field B_0 low enough to couple the principal orbits. It is therefore possible to tilt **B** so as to eliminate breakdown except along a single line and thus create a linear chain. At such values of B that make the small orbits, assumed coherent, resonate the chain is opened and the effective path of the electrons concerned greatly increased. This is something that does not happen with the hexagonal network, and it leads to especially strong oscillations of resistivity in tin. Young[49] has analysed this case in some detail and investigated it experimentally with considerable success, as well as related phenomena including oscillations of the thermoelectric power[50] and Hall coefficient.[48] One result, though lying outside the present scope, is surprising enough to deserve mention – the vanishing of the oscillations of thermal resistivity[50] though not of electrical resistivity, at a certain field strength which is very nearly proportional to temperature. Young shows this to arise from the form of the temperature factor – it is d^2R_T/dX^2, rather than R_T itself, that governs thermal conduction, and $R_T''(X) = 0$ when $X = 1.62$.

144

2. *Magnetic breakdown in aluminium*

Strong slow oscillations of ρ_{xx} are found when **B** lies very close to [100] or [110], and less strong at a few other localized directions. It is not easy to visualize how the orbit network arises from figs. 3.8 and 3.9 but Balcombe and Parker,[51] who first observed the oscillations, and Grossbard,[52] who made a more systematic survey of a high-purity sphere by electrodeless methods, have drawn sections which show the processes involved. In addition to the oscillations the general trend of ρ_{xx} is interesting. For some directions of **B** it saturates almost perfectly while for others, even when the oscillations are weak, a quadratic rise is found which Grossbard is able to explain fairly convincingly in terms of extended orbits resulting from breakdown. For most directions of **B**, however, a slow, almost linear rise is found, much as Fickett[53] observed in a polycrystal, and it is likely enough that the number of crystallites showing a strong quadratic rise was small enough in his sample to produce no significant change in the average linear effect. Grossbard is unable to decide from the available evidence whether this linear effect is to be attributed to breakdown or to small-angle scattering as suggested in the previous chapter.

130

3. Interferometer oscillations

Stark and Friedberg[54] observed and studied in detail very slow oscillations in magnesium, but with **B** normal rather than parallel to the *c*-axis. This is another case where the detailed shape of the Fermi surface is hard to visualize, but the essential feature which they elucidate from it is somewhat unusual, and the resulting resistance oscillation has no counterpart in the de Haas–van Alphen effect. It does not arise from the quantizing of an orbit but from the possibility, as in Young's double-slit interferometer, of a wave (the electron in this case, of course) having two alternative paths between two points. The real trajectories are more complicated than the simplified double junction of fig. 2(*a*), which suffices to illustrate the point. At both high and low field strengths an electron emerges along the same arc as that by which it enters, but at intermediate field strengths it may emerge along either. As the labels on the diagram indicate, the calculation is very easy, giving for P_2 the probability that the electron will switch paths,

$$P_2 = |pq(e^{i\phi_1} + e^{i\phi_2})|^2 = 4PQ\cos^2 \tfrac{1}{2}(\phi_1 - \phi_2), \tag{4.59}$$

where $P = pp^*$, $Q = qq^*$ and $\phi_1 - \phi_2$ is the phase difference between the paths, which is $\mathscr{A}_k/\alpha$ if $\mathscr{A}_k$ is the enclosed area in *k*-space; the probability P_1 of staying on the same path is of course $1 - P_2$. The oscillations, as with Young's slits, are strictly sinusoidal. The complete orbit executed by the electron is quite different for the two possible outcomes and the resistivity consequently also oscillates. The amplitude of the oscillations is only very little dependent on temperature, as with the zone oscillations but not for the same reason. Here the area $\mathscr{A}_k$ is energy-dependent but to a much lesser

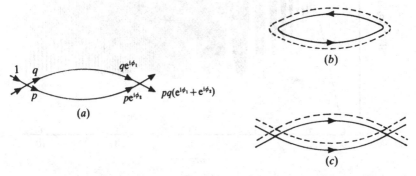

Figure 4.20 Schematic representation of Stark and Friedberg's[54] interferometer network.

extent than a typical orbit area. The explanation is to be seen in figs. 20(*b*) and (*c*), where energy-surface sections are drawn for two neighbouring energies. With the interferometer (*c*) the principal change is a lateral shift and the change in area is governed by the difference between the Fermi velocities on the two arms, which may be much smaller than either separately.

4. Magnetic breakdown and Shoenberg interaction in beryllium

The vigorous oscillations observed in beryllium[55] when *B* lies along the hexad axis are illustrated in fig. 21. In essence they are to be explained in the same way as those in magnesium and zinc. Here, however, the branch of the Fermi surface responsible for the small triangular orbits is larger, so that the oscillations are faster; moreover it is nearly cylindrical and the magnetic oscillations of the de Haas–van Alphen effect are exceptionally large. Conditions are ideal for a strong magnetic (Shoenberg effect) interaction. This has been reviewed in some detail elsewhere[46] and there is little to be gained by giving a resumé of the processes involved, especially as the consequences seem to be unique to beryllium. Unlike the comparably vigorous oscillations in magnesium (fig. 13(*a*)) there is no need to invoke

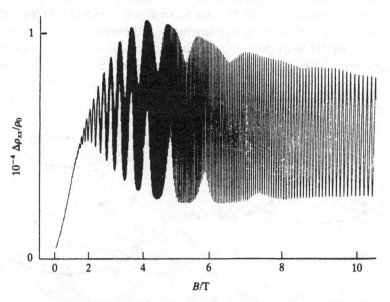

Figure 4.21 Magnetoresistance oscillations in beryllium at 1.39 K (Reed and Condon[55]); note that the field scale is non-linear.

phase coherence round the larger orbits; local coherence round the smaller orbits suffices. For all its remarkable appearance, this example of breakdown oscillations is as well understood as any, though this is not a very proud claim – the whole field of magnetic breakdown has been covered only patchily by experiment and theory, and much remains to be done by those who can see a promising way forward.

5 Inhomogeneous materials

Two types of macroscopic inhomogeneity will concern us – intentional and inadvertent. Typical of the former are polycrystals and of the latter the inevitable flaws, voids, etc. in real samples. We shall also have to consider the significance of many studies of potassium whose extremely strange behaviour must to a large degree be the result of inhomogeneity arising from a phase transition. None of these problems is easy to discuss with confidence, and the whole chapter should be treated as exploratory, certainly not authoritative.

Polycrystals

The resistivity of a random polycrystalline mass must be some sort of average over individual crystallites, but it is no easy matter to determine what average should be taken. If ρ_{ik} varies over a small range only, as the orientation is changed with respect to **B** and $\mathscr{E}$, the choice of averaging procedure will hardly matter. At the other extreme a sparse random distribution of conducting granules in an insulating matrix has an infinite $\langle\rho\rangle$ but a finite $1/\langle\sigma\rangle$ and the details of the averaging process are obviously important.* In this case, as the proportion of space occupied by conducting granules is increased to about 10–20% (the *percolation threshold*) chains of granules in contact make their appearance and there is an abrupt change from the insulating to the conducting state. A similar effect causes a mixture of normal and superconducting domains to lose its resistance before the temperature has been lowered to the point where all regions are superconducting. Problems involving the percolation threshold are extremely difficult to analyse and will not be discussed further. An excellent general survey has been given by Landauer,[1] with copious references to early as well as recent work.

---

*Angular brackets are used in this chapter to denote the average of a quantity over all crystallites, and a bar to denote the measured value for the polycrystalline sample.

Let us start with a rather simple case, the longitudinal magnetoresistance in polycrystalline copper (fig. 1). Calculations for single crystals, along the lines described in chapter 3, lead to the theoretical saturation values of $\Delta\rho_{zz}(\infty)/\rho_0$ shown in fig. 2. The real shape of the Fermi surface was used to determine the range of latitudes swept out by each neck, but otherwise the surface was assumed to be spherical. The average value of $\Delta\rho_{zz}(\infty)/\rho_0$ is 0.78, to be compared with something a little over 1 in the measurements. In view of the discussion in chapter 3 which showed how only a little small-angle scattering sufficed to raise the magnetoresistance, and the slow approach to saturation in fig. 1, characteristic of small-angle scattering, the discrepancy need not be taken seriously. We should ask, however, what justification there is for averaging ρ rather than, say, σ.

The resolution of this question lies in the current-jetting phenomenon discussed in chapter 2. With even the rather modest values of $\omega_c\tau$ in fig. 1

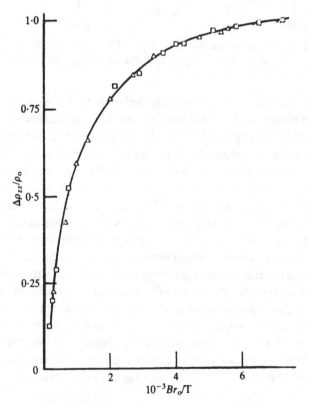

Figure 5.1 Longitudinal magnetoresistance of polycrystalline copper at 4 K (de Launay *et al.*[2]). At the right-hand side $\omega_c\tau \sim 35$.

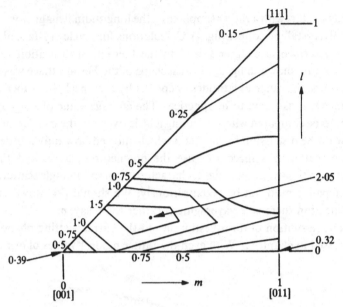

Figure 5.2 Calculated saturation value, $\Delta\rho_{zz}(\infty)/\rho_0$, for a single crystal of copper with **B** and **J** lying along $[l, m, 1]$; the model of fig. 3.27 was used.

currents running parallel to **B** will be little deflected by the expected, relatively small, fluctuations of ρ_{zz}. It should be a good first approximation, especially at higher field strengths, to assume the current density is uniform, so that each crystallite contributes to $\bar{\mathscr{E}}_z$ strictly in proportion to its ρ_{zz}.

When we turn to the transverse magnetoresistance the reasoning is not so clear-cut, but the analogous argument suggests that σ is to be averaged, as was first tentatively proposed by Ziman[3] though without more than pragmatic justification. The following explanation is due to Stachowiak.*
When $\omega_c\tau \gg 1$ the fact that σ_{zz} is very much greater than any other component (the origin of current-jetting) implies that even if the current lines are deflected on meeting an uncongenial crystallite, they will not set up any sizeable component $\mathscr{E}_z$. It is then likely to be a good first approximation to set $\mathscr{E}_z$ and all its derivatives equal to zero. But curl $\mathscr{E} = 0$, and if $\partial\mathscr{E}_z/\partial x$ vanishes, so does $\partial\mathscr{E}_x/\partial z$; similarly $\partial\mathscr{E}_y/\partial z = 0$. If $\mathscr{E}_x$ and $\mathscr{E}_y$ have no z-variation in spite of the inhomogeneity of the polycrystal we may reasonably conclude that inhomogeneity similarly generates no variation with x or y. Given that $\mathscr{E}$ is uniform ($\mathscr{E}_x$, say) the current is everywhere

---

*I cannot find this in any of his published papers, and believe he must have told me of it in conversation.

proportional to the local values of σ_{xx}, σ_{yx} and σ_{zx}, and the average current is to be found by averaging σ_{ij} over all crystal orientations. The behaviour here is similar to that discussed in chapter 2, when the metal contains voids. Discontinuities of J_x at crystallite interfaces result in current jets along the z-direction, fanning out and ultimately averaging to zero at distances about $\omega_c \tau$ times the crystallite dimension.

Ziman, recognising that when $\omega_c \tau \gg 1$ the conductivity in copper is dominated by open and highly extended orbits passing through the necks, considered a metal whose Fermi surface was a set of non-intersecting cylinders and showed that the resistivity of a polycrystal should eventually be proportional to B, as apparently required by the measurements shown in fig. 1.6. To go beyond this and tackle the real Fermi surface is a tedious matter which need not be rehearsed here.[4] It involves fleshing out the skeletal diagram of fig. 3.17 by quantitative estimates of open and extended orbit conductivity as it varies with the direction of **B**. The conclusion (subject to a number of simplifying assumptions, e.g. that $\sigma_{xz} = 0$, and that τ is the same everywhere on the Fermi surface) is that when $\omega_c \tau$ is large,

and
$$\begin{rcases} r_{xx} \equiv \bar{\sigma}_{xx}/\sigma_H = (0.46 + 0.011\,\omega_c\tau) \\ r_{xy} \equiv \bar{\sigma}_{xy}/\sigma_H = (0.90 - 2.0/\omega_c\tau), \end{rcases} \tag{5.1}$$

where σ_H is the value of σ_{xy} for the ideal sphere, ne/B. From these expressions $\bar{\rho}_{xx}$ follows from (1.14) which may be rewritten in the present application:

$$\frac{\bar{\rho}_{xx}}{\rho(0)} = \omega_c\tau r_{xx}/(r_{xx}^2 + r_{xy}^2). \tag{5.2}$$

This is shown as the full line in fig. 3, but not below $\omega_c\tau = 10$, for the approximations in (1) are then untrustworthy. Also shown is the curve of fig. 1.6, and if the agreement seems imperfect one should recollect that 7 samples studied by Martin, Sampsell and Garland,[5] though approximately linear up to $\omega_c\tau = 10$ and beyond, had slopes varying over a factor of 2.4, from 0.75 to 1.8 times the slope predicted by (1). It may also be noted that the longitudinal magnetoresistance of these samples was considerably greater than that shown in fig. 1, indicating the by-now-expected enhancing effect of small-angle scattering in purer samples of copper. No allowance for this has been made in the theory, but the discussion in chapter 3 might suggest that the conductivity of the dominant crystallites, having open or highly extended orbits, is likely to be reduced. The effect on $\bar{\rho}_{xx}$ now depends, as is seen from (5.2), on the relative magnitudes of $\bar{\sigma}_{xx}$ and $\bar{\sigma}_{xy}$. At low fields $\bar{\sigma}_{xx}$ is the larger, and decreasing it increases $\bar{\rho}_{xx}$; but at high fields

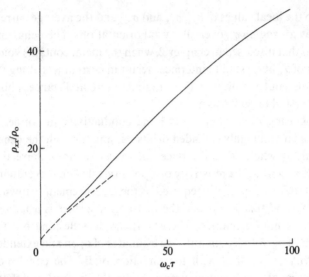

Figure 5.3 The full line shows $\bar{\rho}_{xx}(\omega_c\tau)$ for copper as calculated from (2); the broken line is taken from the experimental curve in fig. 1.6.

$\bar{\sigma}_{xy}$ takes charge, so that $\bar{\sigma}_{xx}$ and $\bar{\rho}_{xx}$ are decreased together by small-angle scattering. In most of the results quoted the former situation obtains (according to (1) $\bar{\sigma}_{xx} = \bar{\sigma}_{xy}$ when $\omega_c\tau = 110$, or $Br_0 = 2.4 \times 10^4$ T) and the measured values of $\bar{\rho}_{xx}$ may be expected to exceed the theoretical values based on catastrophic scattering.

Effective medium theory[1]

The next stage of approximation (beyond which the real difficulties begin) is an attempt at a self-consistent picture of the inhomogeneous medium. It is best illustrated by means of a dielectric analogy, consisting of a random mixture of irregular crystallites, each of which is isotropic but among which there is a wide variation of permittivity, ε. What is the effective permittivity, $\bar{\varepsilon}$, of the mixture? To answer this question, note that in an infinite aggregate of crystallites any chosen one will be found repeated in different places, identical in shape and orientation but surrounded by a different selection of others. Its response to the applied field, when averaged over all possible environments, will (one hopes) be not too different from the response of any one surrounded by the average environment, i.e. a uniform dielectric, $\bar{\varepsilon}$. This response is governed by its own permittivity and by its shape. Here we keep our fingers crossed and trust that an irregular crystallite may be treated as an ellipsoid, for we know that the response of an ellipsoid, when averaged over all orientations with respect to $\mathscr{E}$, is the same as the response of a sphere. By this reasoning we arrive at a practicable recipe, which is to treat each crystallite as a sphere embedded in the uniform matrix of permittivity $\bar{\varepsilon}$. If the field in the matrix at a distance from the crystallite is $\mathscr{E}_0$, the field in the sphere is $\mathscr{E}_{in}$, where:

$$\mathscr{E}_{in} = 3\bar{\varepsilon}\mathscr{E}_0/(\varepsilon + 2\bar{\varepsilon}). \tag{5.3}$$

We know, however, that $\langle\mathscr{E}_{in}\rangle$, the average value of $\mathscr{E}_{in}$ taken over all crystallites, must be the same as $\mathscr{E}_0$, from which it follows that $\bar{\varepsilon}$ must be such that:

$$\langle 3\bar{\varepsilon}/(\varepsilon + 2\bar{\varepsilon})\rangle = 1,$$

or

$$\langle 1/(\varepsilon + 2\bar{\varepsilon})\rangle = 1/3\bar{\varepsilon}. \tag{5.4}$$

The parallelism between permittivity and conductivity allows the analogue 53 of (5.4) to be written unchanged,

$$\langle 1/(\sigma + 2\bar{\sigma})\rangle = 1/3\bar{\sigma}. \tag{5.5}$$

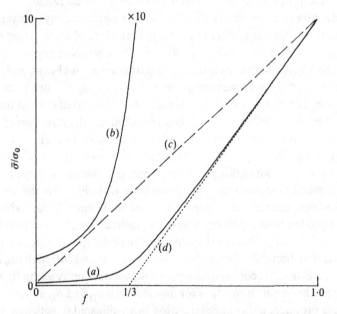

Figure 5.4 (a) Effective medium approximation (5) for the conductivity of a mixture of crystallites in which a fraction f have $\sigma = \sigma_0$, and $1 - f$ have $\sigma = \sigma_0/100$; (b) part of (a) on tenfold enlarged vertical scale; (c) simple average $\langle\sigma\rangle$; (d) solution of (5) when a fraction $1 - f$ are insulating.

As illustration consider what may seem a rather extreme case in which there are two types of crystallites, a fraction f with conductivity σ_0 and a fraction $(1 - f)$ with conductivity $\sigma_0/100$. Then (5.5) is readily solved, to give curves (a) and (b) in fig. 4, markedly different from the straight line (c) which is the weighted arithmetic mean as suggested by Ziman[3] (though he would not have suggested it for this problem!). In the extreme case when the fraction $(1 - f)$ is insulating the solution (5.5) takes the form of two straight lines, $\sigma = 0$ if $f < \frac{1}{3}$ and $\frac{1}{2}(3f - 1)$ if $f > \frac{1}{3}$. There is nothing in the theory to imply that continuous conducting paths should appear when f reaches $\frac{1}{3}$, and no significance should be attached to what appears to be the prediction of a percolation threshold. From an examination of various computer simulations Ziman[6] proposed a simple empirical formula which, applied to a polycrystal in which the average number of facets per crystallite is 13.4 (itself an empirical result), gives the threshold at 11% conducting crystallites, rather different from the 33% resulting from (5). Nevertheless, one has little choice but to conclude that, in default of anything better that might yet be amenable as the basis for computation, the effective medium approximation is well worth employing, though with appropriate scepticism when the property to be averaged varies over a wide range.

It needs, however, considerable extension to apply to, say, polycrystalline copper at high $\omega_c\tau$, where very large variations of σ are expected, depending on the presence or absence of open orbits; moreover σ_{ij} in every crystallite and $\bar{\sigma}_{ij}$ are highly anisotropic. In particular σ_{zz} is always, and by a considerable factor, the largest component so that $\bar{\sigma}_{zz}$ also is much greater than $\bar{\sigma}_{xx}(= \bar{\sigma}_{yy})$ or $\bar{\sigma}_{yx}(= -\bar{\sigma}_{xy})$. As a result the distortions of current flow around a supposedly spherical crystallite spread along the direction of **B** not merely to a distance of one radius but, as we have seen already, to something more like $\omega_c\tau$ radii. They therefore pass through many crystallites and, far from making matters worse, give rise to the hope that the effective medium approximation will work particularly well in this case.

42   In the scaled coordinate system discussed in chapter 2, for which Laplace's equation holds, each crystallite is squashed along the z-direction by a factor $R = (\bar{\sigma}_{zz}/\bar{\sigma}_{xx})^{1/2}$, so that when $\omega_c\tau$ is large the sphere is deformed into a rather thin discus; the depolarizing factor for field and current lying in the plane of the discus is correspondingly small, but near unity along **B**. As noted there, it has been shown that for any tensorial σ_{ij} and $\bar{\sigma}_{ij}$ a uniform applied field produces uniform polarization in an ellipsoidal inclusion. In extending the effective medium theory we may take this for granted, and it only remains to identify the polarization, both its magnitude and direction, by matching boundary conditions at points selected for their convenience.

This task is simplified further if longitudinal–transverse coupling can be neglected, for the polarization of each spherical crystallite and the current flow everywhere lie in the x–y plane, making it necessary to match the boundary conditions at only two points on the rim of a discus. The discussion in chapter 3 provides some justification for the simplification, but doubts must remain when a substantial fraction of the crystallites support open orbits. It may also be noted that unless longitudinal–transverse coupling is neglected the same scaling cannot be applied consistently to the material within the chosen crystallite and to the medium as a whole. This makes a full solution even more troublesome.

The result of the analysis[4] corresponding to (5), but considerably more complicated, takes the form of two equations:

$$\langle [\sigma_{xx}(\bar{\sigma}_{xx} + \lambda\sigma_{yy}) - \lambda\sigma_{xy}(\sigma_{yx} - \bar{\sigma}_{yx})]/M \rangle = 1/(1 + \lambda) \qquad (5.6)$$

and

$$\langle (\lambda\bar{\sigma}_{xy}\sigma_{xx} + \bar{\sigma}_{xx}\sigma_{xy})/M \rangle = \bar{\sigma}_{xy}/(1 + \lambda)\bar{\sigma}_{xx}. \qquad (5.7)$$

In these equation $\lambda = D/(1 - D)$ where D is the depolarizing factor in the plane of a discus whose ratio of diameter to thickness is R:

$$D = [R^2 \tan^{-1}(R^2 - 1)^{1/2}/(R^2 - 1)^{1/2} - 1]/2(R^2 - 1)* \qquad (5.8)$$

and

$$M = (\bar{\sigma}_{xx} + \lambda\sigma_{xx})(\bar{\sigma}_{xx} + \lambda\sigma_{yy}) + \lambda^2(\sigma_{xy} - \bar{\sigma}_{xy})^2. \qquad (5.9)$$

The application of this theory to a metal like copper involves much detailed estimation of how σ_{ij} varies with crystal orientation, but in the end it turns out that $\bar{\rho}_{xx}/\rho_0$ should rise almost proportionally to B, only saturating at a value well in excess of 100. Thus fig. 1.6 represents only the beginning of a long climb. This is rather an accidental result – the prevalence of extended and open orbits happens to be about right. For various reasons the most careful experiments, those of Fickett,[7] cannot be expected to agree very closely with theory. For once, small-angle scattering is not the principal culprit since, as (1.41) shows, if $\Delta\rho/\rho_0 \propto B$, $\Delta\rho(B)$ is independent of ρ_0; this means that a change in the effective value of τ with B does not move the resistivity off its straight line. The obstacles to a valid comparison here are the size of the crystallites and the texture of the polycrystal. In a drawn wire, even after annealing, there are markedly preferred orientations to invalidate the assumption of randomness. Further, to get enough crystallites in any one cross-section, while aiming at high purity and correspondingly large values of $\omega_c\tau$, implies crystallite dimen-

---

* There is an error in equation (3) of the paper where r should be read as r^2, i.e. σ_{zz}/σ_{xx} (r as used there $= R$ here). D is the same as $\varepsilon_0 L'$ in (2.31).

sions comparable to the free path of the electrons and possibly a dominance of boundary scattering; it is then dubious whether the interior of the crystallite can be treated as a homogeneous conductor. Both these objections could probably be evaded by rapid casting, annealing and machining of a large polycrystalline sphere, to be studied by electrodeless methods. But this has not been done.

130   Other metals, including Al[8] and In,[9] have been studied as polycrystals;
131   both show only a rather small increase in ρ_{xx} ($\Delta\rho/\rho_0 \sim 2$ or 3) such that Ziman's proposal to average σ_{ij} is probably an excellent procedure.

Potassium

The simplest of metals has generated the most intricate puzzle that we have to worry about. The first indication of trouble was the observation of a nearly linear increase of resistance up to very high values of $\omega_c\tau$ (see fig. 1.3). Subsequently, the behaviour has been found to be irreproducible and increasingly eccentric. Something very similar might be discerned by the detached observer in the attempts at theoretical explanation. Fairly detailed reviews have been written by Overhauser,[10] by Fletcher[11] (who is inclined to be unsympathetic to Overhauser's interpretation), and by Coulter and Datars[12] (who are more sympathetic). The following account describes how the problem came to be recognized, though without excessive adherence to the historical sequence. The three reviews cited here provide a full bibliography of what is now an extensive field.

Fig. 5 shows examples of the transverse magnetoresistance, measured on 1 m long extruded wires as used originally by Babiskin and Siebenmann.[13] The purity was varied by adding small amounts of sodium. Although the largest value of $\omega_c\tau$ in this diagram is about 80, the linear rise has been found with purer samples to continue up to values of many hundreds. Extrapolation back to zero field as practised by some authors serves to illustrate the irreproducibility but has little other significance, except in so far as it might give an upper limit to the magnetoresistance in an ideal sample, devoid of whatever is responsible for the variable linear term. The least pure sample here, for instance, extrapolates back to a value of 0.6% for $\Delta\rho/\rho$, which is not much more than might be expected for a metal with a very nearly spherical Fermi surface. Babiskin and Siebenmann obtained a similar low value with a rather impure sample. It should be remembered that the long samples contained many crystallites, and it is in the highest degree improbable that they should all have been oriented so as to avoid exciting open orbits, 'hot spots'[15] or other hypothetical disturbances to

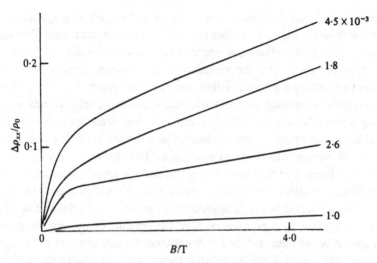

Figure 5.5 Transverse magnetoresistance of potassium wires of different purity (Taub et al.[14]); from the top downwards, $r_0 = 880, 2300, 650$ and 353. The figure against each curve shows the Kohler slope of the high-field line.

what had hitherto been believed to be a nearly isotropic conductor. The explanation that relies on hot spots – localized regions on the Fermi surface where exceptionally strong scattering occurs – is one of the most vulnerable of the theories proposed, in that such spots must be present in all samples irrespective of details of their construction, and should not be able to exert an influence in any that is stronger than their influence in the best behaved. The same objection applies to any proposal such as that of O'Keefe and Goddard[16] for a different electronic structure that could generate open orbits. Unless it can explain why this structure only appears in a fraction of the samples it carries little conviction. It is a measure of Overhauser's ingenuity that he is able to suggest more plausible ways round this difficulty, and (with whatever misgivings) we must respect the determination with which he maintains his explanation of the various phenomena, including their caprices.

Before leaving the curves of fig. 5 let us note that most appear to lack a quadratic variation at the weakest field strengths. The lowest, however, seems to be roughly quadratic up to about 0.2 T; probably more points taken with the purer samples at lower fields would have shown at least rough agreement with Kohler's rule, and a similar quadratic variation for $\omega_c \tau$ less than 0.5. As far as one can judge, the most impure sample reaches three-quarters of its 'saturation' value at about 0.4 T, where $\omega_c \tau = 1.2$. It may be noted that the square model Fermi surface, whose magnetoresistance is shown in fig. 1.19, reaches the same fraction of saturation when $\omega_c\tau$ is just under 1.0. Now this result applies to orbits with four discontinuities of direction per cycle, and the variations of velocity, relaxation time, etc. around any orbit on a Fermi surface with the cubic symmetry of potassium are likely to run through two, and often more, cycles per orbit. It then seems reasonable to believe that this tiny initial magnetoresistance reflects the ideal behaviour of potassium, considered as a simple metal, and that some other explanation, intrinsic or extrinsic, must be sought for everything else.

It is far from clear that a single mechanism will prove adequate, for the form of $\Delta\rho/\rho_0$ varies greatly. All the curves in fig. 5 have a pronounced knee, but others published by the same group are linear almost all the way, so that they extrapolate to a point very near the origin. Again, annealing the samples at room temperature for many days usually changes the shape, reducing the initial slope by a large factor while changing the Kohler slope* much less (the purest and least pure samples suffered reduction of the slope on annealing, while samples of intermediate purity suffered an increase).

No one has offered an explanation that accounts for the initial slope, followed by a transition to a slower linear variation; historically, the most popular explanation for the latter is that it reflects the presence of gas bubbles or voids of some other origin. According to (2.33), when $\omega_c\tau$ exceeds unity, ρ_{xx} increases linearly at a rate $0.486\,f\omega_c\tau$, where f is the volume fraction of spherical voids; ρ_{zz} behaves similarly, but with the rather larger coefficient 0.637.[†] The Hall constant (or ρ_{xy}) has no significant linear increase. Although it is an attractive idea to blame the behaviour on voids, not least because there is some evidence[18] that degassing of the potassium results in a smaller Kohler slope, the volume fraction that must be occupied by voids is rather large, being 0.2% when the slope is 10^{-3}. Measurements of the density[19] and lattice constant,[20] both at room temperature, agree to better than 0.01% and make clear that in carefully prepared samples the void content is far too small to explain the observed Kohler slopes. The question remains, however, whether any of the samples used for resistivity measurements were prepared carefully enough. For his density measurement Stokes[19] distilled potassium *in vacuo*, while the most that others have done is to heat it to somewhat above the melting point *in vacuo*. One would

---

* The Kohler slope is the slope of the linear magnetoresistance curve when $\Delta\rho/\rho_0$ is presented as a function of $\omega_c\tau$.

† Stinson[17] has applied effective medium theory to this problem, but without changing the position significantly.

wish to see magnetoresistance results for Stokes' material, though more recent experiments that we shall come to presently make it exceedingly doubtful whether any test as simple as this would eliminate all problems. Before leaving the subject of voids it is worth pointing out that, like the crystallites in copper, any present are almost certainly smaller than the electron free path. It is not impossible that they exert a stronger influence than continuum theory predicts, but no calculations have been attempted. 186

The Hall constant, interpreted according to (1.52) as $1/ne$, and its tendency to drift at high field strengths, has been frequently invoked as evidence against the free-electron picture of potassium. As already discussed in chapter 2, much of the evidence is suspect, coming as it does from helicoñ measurements on samples with insufficiently rigid mountings, so that the soft-helicon effect depresses the frequency and enhances the apparent value of n. 65

Leaving the more-or-less conventional investigations we turn to more recent electrodeless methods and find ourselves in the thick of apparently irreconcilable observations. We might well echo the words of Blaney[21] in his Ph.D thesis (1961) on magneto-acoustic effects in potassium – 'This work sets out to be a Baedeker and ends as a collection of traveller's tales'. The first to rotate a sphere of monocrystalline potassium (diameter = 22 mm) in a fixed field was Lass,[22] who confirmed the linear magnetoresistance with a rather large Kohler slope, about 10^{-2}. The sphere was spark-machined from an ingot and etched to remove the undoubtedly damaged surface layer (potassium is so soft at room temperature that spark-machining causes flow). Possibly not enough was removed and, as Lass points out, the outer layers play an inordinately large part in determining the response, as witness the fifth power of a in (2.43). All the same, the resistance ratio measured by rotation in a low field was about 5000, which is quite good. At his highest field, 2.2 T, $\omega_c \tau \sim 80$. There was no variation of torque amounting to more than 2% during a complete rotation of the sphere, and this was the case both when the axis of crystal growth was parallel and when it was normal to the rotation axis. This is significant in the light of later results, and it must also be noted that the sphere was supported with minimal constraint. Unfortunately the published account lacks details about the techniques, which are only to be found in the unpublished Ph.D thesis. Also there is no information about crystal orientation in any of the experiments.

The results obtained a year later by Schaefer and Marcus[23] present a strong contrast. Their samples were considerably smaller (diameter between 2 and 7 mm) and were made by solidifying molten globules under paraffin. They sat in the sample holder on a bed of mineral grease that held them firmly on cooling, and also must have set up substantial strains as a result of differential contraction. Almost all out of a large number studied showed strong anisotropy in the torque, with a $\cos 2\theta$ variation at low fields on which was superposed a stronger $\cos 4\theta$ variation at higher fields. This is very much like Lass'[24] expression (2.53) for the torque on a spheroid, and indeed he showed[25] that if b is set equal to 0.87 the magnetoresistance curves at different orientations could be rather well reproduced. He suggested that the dimple resulting from solidification of the liquid sphere was the basic cause of the effect observed.

This cannot be the whole story, however, for Holroyd and Datars[26] also obtained the same pattern of angular variation with larger spheres, grown in moulds, and certainly nearly enough spherical to rule out Lass' explanation. The enormous magnitude of the anisotropy is clear from fig. 6. There is strong evidence that the effect, which is essentially an enhancement of the torque at certain angles rather than a diminution in other directions, occurs only in strained samples such as are partially coated with oil or grease, and therefore subject to inhomogeneous stress when the coating freezes. This would explain why Schaefer and Marcus found the effect, and Holroyd and Datars confirmed it with oil-coated spheres, while Lass and Simpson, who used almost free and clean samples, saw nothing extraordinary.

The discussion of the rotating sphere technique in chapter 2 leaves little room for explanations based on the nearly spherical Fermi surface that is derived from the de Haas–van Alphen effect;[27] and whatever doubts may

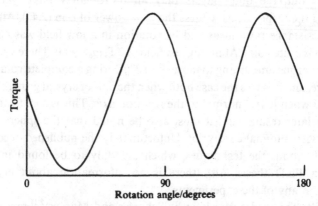

Figure 5.6 Angular variation of torque on a potassium sphere of diameter 17.1 mm, rotating in a field of 2.3 T (Holroyd and Datars[26]).

remain concerning the quantitative interpretation of the measurements they do not extend to the point of attributing everything seen by Datars and his co-workers to artefacts. The only plausible origin so far suggested for the angular variation is that there are open (or at least highly extended) orbits, and this demands a significant change in crystal symmetry. Overhauser[28] has proposed incommensurate charge-density waves as the mechanism, but has been forced by a succession of new experiments to modify the details several times in a way that has not enhanced the credibility of his idea. Nevertheless, his critics have not attacked the theoretical calculation directly but rather pointed out, with understandable vigour, that it does not explain all the facts. It is only right, however, that until something better turns up it should remain on the table.

For our purpose a brief account will suffice. Overhauser's analysis indicates that a lowering of energy can be effected by modulating the lattice spacing so as to produce fairly small energy gaps on planes in reciprocal space that are nearly tangential to the free-electron Fermi surface. A typical charge-density wave then pulls out the Fermi sphere into the shape of a lemon (the distortion is small) with points or very small contact areas on the new planes of energy discontinuity. These planes are thought to lie nearly, but not quite, normal to [011] directions, so that there are 48 different possibilities. The unstrained crystal would consist of a mass of domains, of undetermined size, encompassing all orientations, while the effect of strain would be to give preference to some. From this the appearance of large anisotropy in strained spheres, and none in unstrained, is made possible.

The measurements by Fletcher[29] on longitudinal electrical and thermal magnetoresistance strongly support the attribution of the anomalies to inhomogeneities, whatever they may consist of. This particular model of Overhauser is, however, open to pertinent experimental objections:

1. Careful repetition[30] of de Haas–van Alphen measurements shows no significant angular variation of amplitude. If the sample were a single domain the signal would be lost when the relevant orbit passed through the contact points. On the other hand, if it were a mass of small domains, with each orbit passing through several, there would be a significant loss of amplitude overall and this is not acceptable to experimenters. A similar argument applies to the lack of anisotropy of the Knight shift.[31]

2. Open orbits should produce a profound change in the transverse magnetoresistance but not in the longitudinal. Experiment[18] shows, however, that if anything the longitudinal effect is greater

60

than the transverse. At this point we must begin to think about multiple causation; perhaps the linear magnetoresistance is, after all, to be attributed to voids or other small defects, which we know give a stronger longitudinal effect.

3.   In that case, we expect a defect-free sample to behave very similarly to a free-electron metal, as indeed is found by Taub *et al.*[14] and by Babiskin and Siebenmann[13] in their best samples.

The least objectionable starting-point for interpreting the experiments is that unstrained potassium is a perfectly respectable body-centred cubic metal, but only just. Prepared in a strain-free state, it seems it will behave like an almost perfect free-electron system, but strains can all-too-readily precipitate another modification which exhibits open or extended orbits. Very small (~ 10 μm) spheres embedded in oil give sharp nuclear resonance lines[31] incompatible with density variations, and therefore probably are free from shear deformations, unlike the larger spheres of Schaefer and Marcus.[23] The rolled foils used for electron spin resonance[32], however, showed clear signs in their double peaks of having departed significantly from uniformity. One cannot resist the view that potassium is capable of undergoing a structural transition. What is less acceptable in Overhauser's explanation is his unshakable faith in the precise mechanism he has championed, and insistence that the new charge-density-wave phase is stable in all circumstances at low temperatures.

Ironically, one of the most telling criticisms of Overhauser's position was sparked off by his own analysis[33] of the martensitic phase transformation in lithium. At about 74 K the body-centred cubic crystal transforms into a close-packed structure, and he showed that the neutron diffraction data were well fitted by the 9R arrangement also found in one of the modifications of samarium. Instead of the common close-packed forms, hexagonal with layers in *abab* alternation, or cubic with *abcabc* alternation, the 9R has a repeat pattern of nine layers, *abcbcacab*, and it can be reached from the body-centred form by a comparatively small shear and a little subsequent readjustment. Wilson and de Podesta[34] have pointed out that the satellite reflections in the neutron diffraction pattern of potassium are even better fitted by a 9R phase than by the incommensurate charge-density waves that in Overhauser's eyes they seemed to confirm. It is not suggested that potassium actually undergoes a transformation in its unstrained state, but that the two phases are so nearly equal in free energy that small strains may easily cause a local phase change. It is noteworthy that sodium, which certainly undergoes a shear transformation (though not necessarily into the

9R structure) gives indications in its diffraction pattern of local pretransitional deformations above the critical temperature. Moreover the rotation pattern of a sodium sphere in a transverse field[35] exhibits anisotropy not greatly different from that found with strained potassium spheres.

If the layers in the 9R structure are strictly equidistant there seems no reason why the Fermi surface in the new form of potassium should depart from spherical shape any more than in the old. However there are two different types of layers, two-thirds having the same neighbours (like b in *aba*) and one-third having different neighbours (like b in *abc*). One may not then assume that the spacings of the two sandwiches will be identical; if they are not, extra energy gaps will appear within the Brillouin zone of the simple closed-packed form, dissecting the Fermi surface by parallel planes and giving rise to the possibility of open orbits. No calculations have been carried out yet to estimate the likely size of the effect, if it occurs, but the 9R hypothesis must for the present be kept in mind, along with Overhauser's original charge-density wave, as contenders in the struggle to explain the anomalies of magnetoresistance.

So far no mention has been made of the extraordinary high-field studies of Coulter and Datars,[36] of which an example is shown in fig. 5.7, being the torque on a sphere rotating in a field of 8 T. If one did not know the exemplary record as an experimenter of the senior author one would be

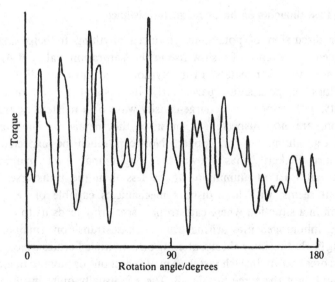

Figure 5.7 Angular variation of torque on a potassium sphere of diameter 4 mm, rotating in a field of 8.5 T (Coulter and Datars[36]).

tempted to dismiss the results as showing the defective nature of the new procedure used here. They have, however, presented in a later account[12] a reasoned defence of the equipment and, supported by Overhauser,[10] claim that the behaviour is compatible with a domain structure of differently oriented charge-density waves, producing a wide variety of open-orbit directions in different regions. It remains to be seen whether all the observations, including a strong temperature dependence of the pattern of peaks even below 1.4 K, can be accounted for by this model or by the 9R model. In view of the observations of de Podesta and Springford[37] on helicons in spherical single crystals at 8 T, which are free from such exaggerated eccentricities, one must still hold to the view that there is such a metal as normal potassium, but it is easily induced to develop abnormalities. There is no point in discussing the matter further in a book devoted to the phenomenon of magnetoresistance. Let us be content for so long as we can believe that unstrained potassium comes very close to the ideal free-electron metal, and leave the problems of strained potassium to the materials scientist in the first place. He could do worse than repeat, much more systematically than hitherto, the experiments that have given such bizarre results, but with lithium and sodium, which are known to suffer transformation, instead of with potassium. Rubidium also, which hovers on the verge of lattice instability[40], is worth further study.

Last thoughts on linear magnetoresistance

Has our discussion of potassium produced anything to help explain the frequent occurrence of a slow rise in the 'saturation' value of $\Delta\rho/\rho_0$ shown by many other metals? Probably not, or at most very little. The peculiarities of potassium, particularly its instability, or imminent instability, with respect to a charge-density wave or a martensitic transformation, are not suspected in most of the metals we have considered, e.g. aluminium or copper. Nor can much be expected of voids, since Fickett[8] has shown the longitudinal linear magnetoresistance, at least in aluminium, to be less than the transverse. We are left, it seems, with four distinct mechanisms capable of delaying saturation in a situation where catastrophic scattering leads us to expect it – voids, inhomogeneities arising from a phase transition, small-angle scattering (including inter-sheet scattering on a multiple Fermi surface) and magnetic breakdown. In each metal we may expect one or more to operate, but certainly not the same mix in all. There is usually quite insufficient evidence in any particular case to allow a firm decision, and indeed when, as

53
61
130
174

in potassium, the body of evidence is large and very varied, the decision seems to be just as far from conclusive. Finally one must add that Mahan,[38] in a long and dense analysis, concludes that many-body effects (which were ignored in the treatments discussed in chapter 4) lead to modifications of the scattering process in strong magnetic fields, and thence to linear magnetoresistance. It is difficult to believe that this mechanism will account for the high variability of the Kohler slope in potassium; nor is it easy to disentangle it from the other, and probably dominant, effects already noted in other metals. It may, however, provide an explanation for recently investigated magnetoresistance in heavy-fermion metals.[39]

6   Size effects

When measurements are made on thin plates or on wires of small diameter, there is almost always extra scattering of electrons by the surface which enhances the resistance, especially if the mean free path in the bulk material is comparable to the smallest dimension of the sample. A considerable early literature on the subject is well summarized by Chambers.[1] In itself the size effect when $B = 0$ would not concern us but for the fact that in deriving magnetoresistance plots the value of $\Delta\rho/\rho_0$ is sensitive to what is assumed about the zero-field resistivity. We shall begin, therefore, with a brief survey of the problems that arise when one tries to estimate the size-effect correction for ρ_0. The theory, at first sight, seems to be not so much intrinsically difficult as rather tedious, but this is to some degree an illusion, as the lurking presence of small-angle scattering introduces considerable complications. When a magnetic field is applied the difficulties become manifest, and our next task must be to obtain enough understanding to appreciate the variety of experimental behaviour, which includes negative and oscillatory magnetoresistance. Lastly we shall explore the effects that arise when special dispositions of electrodes and sample shapes are chosen. In all these cases the sample is truly three-dimensional, even if one or more dimension is highly restricted. Of recent years interest has concentrated more on experiments with two-dimensional conductors, such as are created at the surface of a semiconductor when an electric field confines the electrons to a very thin layer; quantization then eliminates the degree of freedom normal to the surface. This is so different in kind from previously investigated size effects, and so much still a developing field, that a survey written now would be out of date before it was printed. No attempt will be made to cover this topic, even in bare outline, and a list of recent references[2] must suffice.

Size effects without a magnetic field

The size effect is exhibited in its purest form when a wire is used rather than a plate or film, and when the diameter is so much less than the bulk mean free path that one may assume it is only collisions with the rough surface of the wire that interrupt the electron trajectories. The calculation of the resistance may be carried through by analogy with the classical calculation for a rarefied gas in a thin tube (Knudsen flow).[3] In this case gas molecules travel in straight lines between wall collisions which are usually assumed to be diffuse, so that the emergent trajectory is unrelated to the direction of incidence. The effective mean free path in a tube of circular cross-section can then be shown to be just the diameter of the tube. So too in a wire of free-electron metal, under the same assumption of diffuse scattering, l may be taken to be the diameter d, and the effective resistivity is

$$\rho_{\text{eff}} = (\rho_0 l)/d. \tag{6.1}$$

The quantity $(\rho_0 l)$ is enclosed in brackets to indicate it is an intrinsic property of the metal, independent of l, being $\hbar k_F/ne^2$, or $(3\pi^2)^{1/3}\hbar/e^2 n^{2/3}$, for a free-electron metal. It follows from (1) that the resistance per unit length of wire is $4\,(\rho_0 l)/\pi d^3$. There is experimental evidence, which we shall discuss later, that in some special circumstances a certain amount of specular reflection may occur, but normally the assumption of diffusiveness is likely to be justified.

In practice there will be scattering within the bulk of the wire as well as at the surface, and if this is catastrophic and described by a bulk free path l, the measured resistivity will be a compromise between (1) when $d \ll l$ and the bulk value when $d \gg l$. The interpolation formula of Nordheim[4] gives the right answer at both extremes,

$$\rho_{\text{eff}} = (\rho_0 l)(1/l + 1/d), \tag{6.2}$$

and is, as Dingle's[5] detailed calculation shows, nearly enough correct to be useful for all ratios of l/d. His computations indicate that the worst error is a little over 5%, occurring when l is about $5d$.

Looking at the problem from the point of view of the effective path method, one sees that an impulsive electric field, parallel to the wire, excites the same pattern of extra electrons everywhere, but their effective path depends on where in the cross-section they start. Near the surface, half suffer a collision with the surface almost immediately, while of those starting on the centre line even the least favoured travel at least one radius unless they are scattered internally before they reach the surface. When

198   *Size effects*

$l \gg d$, electrons moving nearly parallel to the axis have a much greater effective path than the rest; without internal scattering a fraction $d^2/16x^2$ travel at least as far as x before reaching the surface ($x \gg d$). Thus a fraction $(d^2/8x^3)\delta x$ have effective path between x and $x + \delta x$, contributing to the current an amount proportional to $\delta x/x^2$. The integral determining the current is convergent as $x \to \infty$, so that these few electrons play an important, but not dominant, role. In particular there is no objection to assuming l to be infinite in deriving (1).

The same is not true for conduction in a film of thickness d, if l is infinite. Here a fraction $d/2x$ travel at least as far as $x(x \gg d)$ and the number with effective path between x and $x + \delta x$ is proportional to $\delta x/x^2$, making a contribution to the current proportional to $\delta x/x$. On integration a logarithmic divergence appears as $x \to \infty$, and the contribution of electrons moving nearly parallel to the film is so overwhelmingly important that their internal scattering must not be ignored. The theory for catastrophic scattering was given by Fuchs;[6] in the limit $l \gg d$.

$$\rho_{eff} \sim (4\rho_0 l/3d)/[\ln(l/d) + 0.423], \tag{6.3}$$

the logarithmic divergence as $l \to \infty$ being shown explicitly.

Both (2) and (3) illustrate how $(\rho_0 l)$ serves as a scale factor determining the magnitude of ρ_{eff} when d is known (the denominator of (3) is a slowly-varying function which can be found by successive approximation). As (1.39) shows, in a cubic metal or after suitable averaging over all crystal directions in a non-cubic metal, $\rho_0 l = 12\pi^3\hbar/e^2 S$; only the Fermi surface area distinguishes one metal from another for this purpose. Without troubling oneself over the appropriate average and whether it is the same (as it probably is not) as the average measured in a polycrystalline wire or film, one can make a rough check on the consistency of theory and experiment by using metals for which S is fairly well known. In Fickett's[7] thorough study of polycrystalline aluminium wires of very high intrinsic conductivity he chooses, from a range (3.5–8.8) $10^{-16}\,\Omega m^2$, the value $\rho_0 l = 7 \times 10^{-16}\,\Omega m^2$ to make his size corrections. Now aluminium, apart from the dissection and rearrangement of the Fermi sphere by Brillouin zone boundaries, approximates rather well to a free-electron metal with three electrons per atom, and this would have $\rho_0 l$ equal to $4 \times 10^{-16}\,\Omega m^2$. Førsvoll and Holwech[8] have estimated that Ashcroft's Fermi surface model has lost enough area to raise the coefficient to 5.1, but 7 is most improbably high value. The value of 5.1 is close to the value for a free-electron sphere holding two electrons/atom. The major part of the Fermi surface in aluminium is a hole surface nearly filling the Brillouin zone, so that it may be a good approximation for many calculations to use the 2-electron sphere. They themselves made measurements on plates and films of various degrees of purity, concluding that the experimental coefficient rises from about 6 when the resistance ratio r_0 is 7000 to 8.8 when it is 26 500. These are only isolated examples to show the variability of the value of $\rho_0 l$ that must be chosen to fit size-effect measurements. Brandli and Olsen,[9] in a very detailed review, list 267 references to measurements on 26 elementary metals, which confirm both the variability and the tendency for the measured $\rho_0 l$ to exceed anything that might be expected when S is known from theory or independent measurements. In particular, Fawcett's[10] study of aluminium by means of the anomalous skin effect, while itself liable to interpretative pitfalls, leads him to suggest that S is little different from the free-electron value.

The measured value of $\rho_0 l$ is sensitive to temperature, increasing as the sample is heated. Thus Olsen[11] has found that with pure indium the resistance of a wire of diameter 0.06 mm increased between 3.5 K and 4.2 K twice as much as for one of diameter 2 mm, and a very similar result was obtained by Blatt, Burmester and LaRoy.[12] Since it is natural to attribute the change to the enhanced phonon scattering at the higher temperature one is strongly inclined to invoke small-angle scattering, even at the lowest temperatures, as an explanation for the high values of $\rho_0 l$. The variability of $\rho_0 l$ may reflect genuine differences between samples of the amount of small-angle scattering, and certainly the result of Førsvoll and Holwech just quoted accords with this view, in that one expects it to have less effect when catastrophic scattering is greater.

It may be concluded, though it is not our present concern, that zero-field size effects are an unpromising indicator of the area of the Fermi surface. If this is still a topic of interest, the anomalous skin effect is probably a more reliable technique. It may well be considerably less sensitive to small-angle scattering, though the question has not been studied seriously. Fawcett's results on aluminium, as well as those of Pippard[13] on copper and Morton[14] on silver, all support this view. In the matter of magnetoresistance studies with wires and plates, however, one must be dubious about any size-effect correction unless, exceptionally, there is independent evidence for its magnitude. So long as the correction itself is not very great there is no cause for anxiety, especially as available theories that might be deployed to explain magnetoresistance measurements are subject to enough uncertainty to swamp errors in the correction. Thus Fickett[7] used wires of about 1 mm diameter in which one of the larger corrections he applied, amounting to 30%, might just as well have been 20% or 40%; but since $\Delta\rho/\rho_0$ exceeded 6 in this sample an uncertainty of 10% in ρ_0 is unlikely to matter.

At this point it is worth reverting critically to the assumption that the surface is rough enough to cause isotropic scattering. If it were not so, and some electrons were to retain a memory of their previous velocity along the direction of current flow, the measured resistance would be less. As we have seen, the tendency is rather the opposite, so that there is little motive to abandon the simplest hypothesis, that of random scattering, especially as it would normally require atomically smooth surfaces to achieve specularity. The work of Tsoi and others relating to this point will be discussed later. In bismuth and in semiconductors, or any material in which the Fermi wavelength is much greater than a typical atomic spacing, disorder or roughness on an atomic scale may well be of only minor importance, and partially specular reflection is a real possibility. Measurements by Friedman and Koenig[15] on electropolished plates of bismuth provide a measure of support for this view, the resistance being sometimes rather smaller than might be expected with random scattering. These experiments, however, and the related discussion, are not entirely convincing, with some obvious loose ends remaining. At all events, there is nothing here to compel a reassessment of the view that in most metals random surface scattering is as good an assumption as any. Only with surfaces prepared exceptionally carefully, as in the work of Mitrjaev *et al.*[16] on tungsten, is there evidence to the contrary. The claim by Chopra[17] to have found considerable specular reflection in evaporated films of silver, especially if they are epitaxially deposited on mica substrates, may not be without foundation but the reasoning that leads him to that conclusion is far from convincing. Some support is provided by the optical studies of Bennett and Bennett,[18] but it would take a specialist with experience comparable to theirs to assess how reliable are their determinations of reflection coefficients higher than 99%.

Size effects in the presence of a magnetic field; B ∥ J.

The negative magnetoresistance exhibited in fig. 1.8 is not perhaps very striking, but it is the first observation of this class of effects, by MacDonald,[19] who provided the correct physical interpretation. The simplest case to take is a wire with **B** applied along its length. Suppose there to be no internal scattering so that when **B** = 0 the resistance is given by (1). As soon as **B** is applied some of the electrons, moving nearly along the wire, will have their trajectories wound into a helix that fails to hit the surface. At once there appears a non-vanishing number of electrons which have infinite effective path, and which carry any injected current without generating an electric field. Thus when l is infinite the resistivity in a longitudinal field should vanish however small $\mathbf{B}$ may be. Of course, the number involved is a tiny fraction of the total when $\mathbf{B}$ is small; as soon as the introduction of appreciable internal scattering eliminates their perfect conductivity and allows the other electrons to play a part, the resistance will be very little different from its value in zero field. At higher fields, however, the trajectories of all electrons may be condensed into helices which are much tighter than the wire diameter, and there will then be a central core to the wire within which the electrons are secluded from the possibility of surface scattering. As $\mathbf{B}$ is increased and the core expands until it nearly fills the wire, the effective resistivity tends towards the bulk value.

The formal theory for catastrophic scattering is not difficult to write down. For example, if one chooses the effective path method one must consider how far every electron will travel, on the average, after being created by an impulsive field. So far as generation is concerned, all points are equivalent but L_z, the component of $\mathbf{L}$ along the wire, depends on the distance from the axis of the point of generation. Some awkward geometry is involved in deciding whether, and after how long a journey, a given electron will hit the surface if it is not first scattered. If that distance, measured along the trajectory, is x it is easily shown that the average effect of internal scattering is to replace this by $l(1 - e^{-x/l})$. It would be no great matter nowadays to write a computer programme to do all this, but at the time the theory was worked out by Chambers[20] no such aid was available, and he exercised considerable ingenuity to reduce the computational labour. Since his procedures elucidated the physics clearly and were, incidentally, responsible for initiating the use of trajectory methods as a more easily handled substitute for solving the Boltzmann equation, the effort was far from wasted. But the relevance to experiment is somewhat slender as so few measurements have been made with this geometry, and most of these with metals that are far from free electron-like.

Among the earliest measurements are those of Friedman and Koenig[15] with a single crystal plate, not a wire, of very pure bismuth; they show how a field of about 10^{-2} T suffices in this rather special material to reduce the resistance to less than half its field-free value. The free-electron theory for the plane geometry has been given by Kao.[21] Apart from a small initial increase in resistance, which has not been unambiguously observed, the predicted behaviour is similar to Chambers' calculation for a wire. In gallium, which may be made extremely pure ($l > 1$ cm), Yaqub and

Cochran,[22] using square wires, have revealed an even stronger effect than is found in bismuth. The initial rise of resistance that they find they attribute to the intrinsic magnetoresistance of gallium. Lutes and Clayton[23] made measurements on a number of aluminium wires with l/d values up to 2.5, obtaining comparable reductions in general agreement with Chambers' theory, though with rather large fluctuations in the parameters needed to make a good fit. The Fermi surfaces of bismuth and aluminium are far from spherical, and gallium is even more complicated, so that with all these measurements one must doubt the value of quantitative comparison with theory.

Experimentally, a transverse field is considerably easier to apply than a longitudinal; the sample may take the form of a long wire wound into a tight helix, with **B** parallel to the axis. Conversely the theory is much more difficult, though not quite so hard for a plate as for a wire. With a plate the result depends markedly on whether the transverse field lies in the plate or normal to it; with a wire no such distinction exists and the experimental behaviour shows features characteristic of both arrangements. For a plate made of a free-electron metal the theory is tractable provided l/d is so large that internal collisions are negligible. The presence of any magnetic field, however small, that is not parallel to the current distorts those electron trajectories which, lying in the plane, were responsible for a logarithmic divergence. Neglecting internal collisions does not then preclude a useful theory of the behaviour.

Transverse magnetoresistance with B lying in the plate

We shall consider the two orientations of **B** in some detail, starting with **B** in the plane of the plate, which is taken to lie normal to the y-axis, so that **B** lies along z as usual and **J** along x. A Hall field $\mathscr{E}_y$ is to be expected in addition to the resistive field $\mathscr{E}_x$, but for a free-electron gas $\mathscr{E}_z = 0$. A considerable simplification results from assuming a two-dimensional system, or cylindrical Fermi surface with no z-component of electron motion, and the result obtained is surprisingly like that for a spherical surface, while avoiding the loss of clarity that accompanies the need to integrate over the third dimension. It is also helpful, to avoid mistakes with signs, to assume the charge carriers to be holes; that is, to take e as positive. This affects the sign of the Hall field but of course not that of the resistance.

We address ourselves initially to the low-field behaviour, when **B** is small enough for the orbit diameter $2R = 2k_F/\alpha$ to be greater than the plate thickness d, so that every trajectory starts and finishes on one surface or the

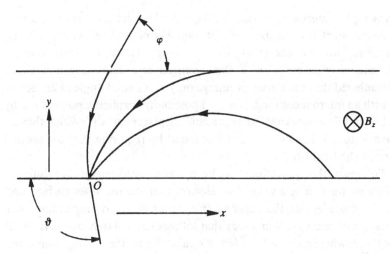

Figure 6.1 Orbits of a free positive hole in a thin plate; the angles ϑ and φ refer to the orbit on the left.

other; i.e. $\beta \equiv d/R < 2$. In fig. 1 the two different types of trajectory are shown reaching the lower surface at the origin of coordinates O; the grazing orbit which separates them is also shown. We shall be concerned only with the electrons as they reach and leave the surfaces, where the potential is $\pm V_0 - \mathscr{E}_x x$. It is easily shown that

$$\cos \varphi - \cos \vartheta = \beta, \tag{6.4}$$

and that the orbit that spans the plate leaves the upper surface at x_0, where

$$x_0 = R(\sin \vartheta - \sin \varphi). \tag{6.5}$$

If, however, the orbit both starts and finishes on the lower surface, it starts at x_1, where:

$$x_1 = 2R \sin \vartheta. \tag{6.6}$$

For most of the calculations in this chapter the effective path method is not as useful as another variant of transport theory, applied as follows to a free-electron metal. Anywhere in the metal the departure from equilibrium of the electron distribution is specified by the displacement, measured by an energy ε, of the Fermi level at each point on the Fermi surface. To avoid space-charge concentrations ε, averaged over the Fermi surface, must vanish. An electron that travels from A to B without intervening collisions starts with total energy (relative to a fiducial E_F) of $\varepsilon_A + eV_A$, and arrives with the same energy, which is now $\varepsilon_B + eV_B$. By following trajectories one may thus relate the variations of ε at different places in the metal. When catastrophic internal scattering occurs, it may be assumed that the scattered electrons start on the local equilibrium Fermi surface, with $\varepsilon = 0$ and therefore with total energy eV. Every trajectory is taken to start from the last catastrophic collision. If this occurred at a boundary it must be remembered that the electrons emerge only into a solid angle of 2π, not 4π as with an internal collision. It is then possible (as explicitly pointed out by Azbel' and Peschanskii)[24] that the same displacement with which they all leave is not eV but $eV + \varepsilon'$; and ε' is found by requiring that the normal current shall vanish.

To apply this procedure to the present problem without internal collisions, we assume a value ε' for electrons leaving the lower surface, and $-\varepsilon'$ for those leaving the upper surface. As for those arriving at the lower surface, ε depends on ϑ in a way that follows immediately from (4)–(6). If $0 < \vartheta < \psi$, where $\cos\psi = 1 - \beta$ (ψ is the value of ϑ for the grazing orbit in the diagram) electrons leave the lower surface with kinetic energy ε' and regain it with $\varepsilon' - e\mathscr{E}_x x_1$, having suffered a potential energy increase of $e\mathscr{E}_x x_1$. If, however, $\psi < \vartheta < \pi$ they come from the upper surface and arrive with kinetic energy $-\varepsilon' - 2eV_0 - e\mathscr{E}_x x_0$. Hence we write for $\varepsilon(\vartheta)$:

$$\left.\begin{array}{ll} \varepsilon(\vartheta) = \varepsilon' - 2e\mathscr{E}_x R \sin\vartheta & ; \quad 0 < \vartheta < \psi \\ \quad -\varepsilon' - 2eV_0 - e\mathscr{E}_x R(\sin\vartheta - \sin\varphi) & ; \quad \psi < \vartheta < \pi \\ \quad \varepsilon' & ; \quad \pi < \vartheta < 2\pi \end{array}\right\} \quad (6.7)$$

In order to determine ε' and $\mathscr{E}_x$ in terms of V_0 we require that there shall be charge neutrality at the edge and that J_y shall also vanish at the edge. If the latter is achieved J_y automatically vanishes everywhere, but internal charge neutrality is not automatic. However, the local value of V adjusts itself everywhere to shrink or expand the Fermi surface uniformly until neutrality is reached. It will be seen that for this problem we do not need to know V, though it will be evaluated in one special case for interest's sake. For charge neutrality $\oint \varepsilon \, d\vartheta = 0$, and on carrying out the integrations we find

$$2\psi\varepsilon' - RF_1 e\mathscr{E}_x - 2(\pi - \psi)eV_0 = 0, \quad (6.8)$$

where $F_1 = 2 + \beta - \int_\psi^\pi \sin\varphi \, d\vartheta$.

To make J_y vanish we must have $\oint \varepsilon \sin\vartheta \, d\vartheta = 0$, and this leads to

$$2(2 - \beta)(\varepsilon' + eV_0) + RF_2 e\mathscr{E}_x = 0, \quad (6.9)$$

where $F_2 = \psi - \frac{1}{2}\sin 2\psi$.

Solving (8) and (9), we have

$$\varepsilon' = (\pi F_2/D - 1)eV_0 \quad \text{and} \quad \text{Re}\mathscr{E}_x = -2\pi(2 - \beta)eV_0/D, \quad (6.10)$$

where $D = (2 - \beta)F_1 + \psi F_2$.

Finally we must determine the total current to obtain the resistance. There is no need to enquire into the details of $J_x(y)$ since consideration of the momentum balance leads directly to the answer. At this point we take note of the caveat mentioned in chapter 1 – since the distribution function is different on the two surfaces there is a net convection of momentum into the plate. Consider the convection of the y-component of momentum, P_y, across unit area normal to y. Associated with the displacement of the Fermi surface by $\varepsilon(\vartheta)$ there is an excess number of carriers $n\varepsilon\delta\vartheta/2\pi E_F$ due to the element $\delta\vartheta$, each having momentum $p_y = mv_F \sin\vartheta$ and moving with y-velocity $v_F \sin\vartheta$. These electrons convect in a positive direction to increase the momentum above unit area at the rate

$$\dot{P}_y = \oint (n\varepsilon\mathrm{d}\vartheta/2\pi E_F) mv_F^2 \sin\vartheta = (n/\pi) \oint \varepsilon \sin^2\vartheta\,\mathrm{d}\vartheta. \tag{6.11}$$

At the lower surface where ε is given by (7), (11) describes the gain of positive momentum by unit length of the plate, while at the upper surface the sign of ε is reversed and there is an equal gain there. At the same time the Hall field increases P_y by $2neV_0$, irrespective of the potential distribution across the plate, and the Lorentz force increases it by $B_z I_x$, where I_x is the total current. In the steady state the net change is zero, so that:

$$B_z I_x + 2neV_0 + (2n/\pi) \oint \varepsilon \sin^2\vartheta\,\mathrm{d}\vartheta = 0. \tag{6.12}$$

Evaluating the integrals as far as possible in terms of elementary functions, and writing for the effective resistivity $\rho_{\mathrm{eff}} = \mathscr{E}_x \mathrm{d}/I_x$, one finds, by use of (10),

$$\rho_{\mathrm{eff}} = Q\rho_0 l/d, \tag{6.13}$$

where

$$Q = \frac{1}{2}\pi\beta^2 \Bigg/ \left\{ 2 - \cos\psi + \frac{1}{3}\cos^3\psi + F_2^2/2(2-\beta) - \int_\psi^z \sin\varphi \sin^2\vartheta\,\mathrm{d}\vartheta \right\}.$$

For the convenient computation of the integral a simple change of variables converts it to the form $(1 - \frac{1}{4}\beta^2)(2-\beta)\int_0^1 [(1-z^2)(1-cz^2)]^{1/2}\,\mathrm{d}z$, where $c = (2-\beta)^{1/2}/(2+\beta)^{1/2}$. The variation of ρ_{eff} (as measured by Q) with B (as measured by β) is shown as curve (a) in fig. 2. Curve (b) was computed by Ditlefsen and Lothe[25] for a plate in which internal scattering gave a value of 36 for l/d. In accordance with (3), (b) starts when $\beta = 0$ at $Q = 0.33$, while (a) starts at the origin and jumps up extremely abruptly. This is to be expected from the earlier discussion, for any B, however small, removes from the action those parallel-moving electrons that had dominated the conduction. Once internal scattering is introduced, their effect is less pronounced and no vertical tangent can be expected at the origin. Indeed, as other computations by Ditlefsen and Lothe make clear, when l/d is

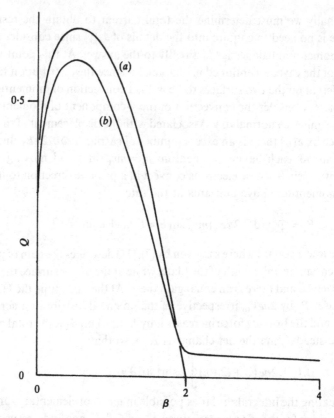

Figure 6.2 $\rho_{\text{eff}}(\beta)$ as measured by Q, (a) from (13) with no internal scattering, (b) for free-electron model with $l/d = 36$ (Ditlefsen and Lothe[35]).

smaller the peak of curves like (b) remains little changed, but the initial value of Q is larger and the initial gradient less abrupt. At the high-field end, when l is infinite Q drops sharply, though with finite gradient, to zero at $\beta = 2$, at the point where an orbit can just fit into the thickness of the plate. With finite l the point $\beta = 2$ is still marked by a rapid flattening off; as β increases ρ_{eff} slowly falls to ρ_0.

Perhaps the most surprising aspect of fig. 2 is the great similarity of the two curves, even though (a) is computed for a two-dimensional model and (b) for a three-dimensional. It appears that the electrons round the equatorial xy-plane of the Fermi sphere control the conduction process. The work of Ditlefsen and Lothe clearly must have involved much analysis and computation, but unfortunately is described so briefly that much of what they did is hard to reconstruct. One cannot doubt, however, that it is essentially correct within the limitations of the free-electron model and catastrophic internal scattering.

Let us look at the distribution of $V(y)$ and $J_x(y)$ across the plate of fig. 1 in the case $\beta = 2$, where it is most easily evaluated, $\mathscr{E}_x$ having fallen to zero. If we draw the orbit that just grazes both surfaces and passes through y at an angle $\vartheta_0(y)$ it is easily seen that every electron passing through y in such a direction that $-\vartheta_0 < \vartheta < \vartheta_0$ has come from the lower surface and lies within a kinetic energy of $e(V_0 - V(y))$, while the rest come from the lower edge and lie within a kinetic energy of $-e(V_0 + V(y))$. Charge neutrality demands that the weighted mean shall vanish, so that:

$$\vartheta_0(V_0 - V) - (\pi - \vartheta_0)(V_0 + V) = 0,$$

and therefore

$$V/V_0 = 2\vartheta_0/\pi - 1 = (2/\pi)\sin^{-1}(1 - 2y/d). \tag{6.14}$$

Given the deformed shape of the Fermi surface at y, J_x can readily be calculated:

$$J_x(y) = (16neV_0/\pi Bd)(y/d - y^2/d^2)^{1/2}. \tag{6.15}$$

Fig. 3 shows V and J_x; since $\mathscr{E}_y = -\partial V/\partial y$ the Hall field rises to ∞ at the sides so that there is something like a double-layer of charge just inside, and space charge everywhere except on the centre plane. The current falls to

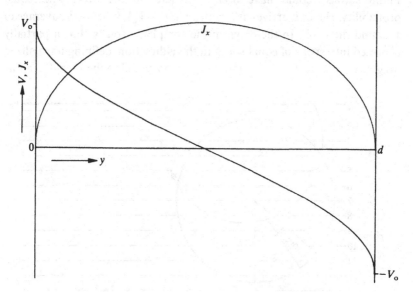

Figure 6.3 Variation of V and J_x across the plate when $\beta = 2$.

zero at the sides, despite the infinite field, since virtually all electrons just inside have come from the same surface and the kinetic energy is independent of ϑ.

The singular behaviour of V at the surfaces continues into the resistanceless domain $\beta > 2$ and makes an analytical determination of the potential distribution, when $R < d$, very difficult, perhaps impossible. It is, however, no great matter to compute $V(y)$ for any choice of β. The basic theory is worth writing down, and for simplicity we shall take $\beta > 4$. Every electron passing through a given point y has then either come from the nearer surface or moves in an orbit that misses both surfaces; taking β as large as this avoids the small extra complication that electrons may have come from either surface. In fig. 4 the y-coordinate is divided into an odd number N of equal intervals, and $V(y)$ is to be found at each of the numbered levels, e.g. at P on level 6. The electrons have diameter equal to m spacings, m being odd (11 in the diagram); $\beta = 2N/m$. Some that pass through P have started on the top surface like that shown, which is centred on level 3; these start with an undisplaced Fermi surface, since $\varepsilon' = 0$ when $\beta > 2$, and arrive at P with $\varepsilon = e(V_0 - V_6)$. But others, such as that centred on level 7, do not touch the surface and require different treatment. If one allows a very small amount of internal scattering, so that $\omega_c \tau \gg 1$, it is clear that the last collision, from which the electron started with an undisplaced Fermi surface, could have been anywhere on the orbit with equal probability. Hence it arrives at P with $\varepsilon = e(\bar{V} - V_6)$, $\bar{V}$ being the average of V round the orbit. In the diagram the complete orbit is shown partially dissected into strips of equal width in the y-direction, defining normalized weights w_1, w_2, etc. for the various arcs, proportional to their length; in this

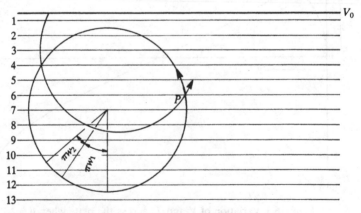

Figure 6.4 Computation of $V(y)$ when $\beta > 4$.

case,

$$\bar{V} = w_1 V_{12} + w_2 V_{11} + w_3 V_{10} + \cdots.$$

Proceeding in the same way one may write $\bar{V}$, and hence ε, for each orbit through P that is centred on one of the discrete levels, and hence write an expression for $\bar{\varepsilon}$ at P. In doing so one must remember that the orbits do not divide the range of directions at P into equal segments, and that the weights w_i must be brought into the average. Finally one writes the conditions that $\bar{\varepsilon} = 0$ at each level, to preserve neutrality, and when everything is sorted out one is left with a set of $\frac{1}{2}(N-1)$ linear equations of the form

$$(A_{jk} - \delta_{jk})V_k + W_j = 0 \quad ; \quad 1 \leqslant j, k \leqslant \tfrac{1}{2}(N-1), \tag{6.16}$$

in which

$$A_{jk} = \sum_{i=1}^{j} w_i(w_{k-j+1} - w_{N-k-j+1+i}), \tag{6.17}$$

and

$$W_j = 1 - \sum_{i=1}^{j} w_i. \tag{6.18}$$

If $i > \frac{1}{2}(m+1)$, $w_i = w_{m+1-i}$ and $w_i = 0$ unless $1 \leqslant i \leqslant m$. The appearance of two terms in (17) reflects the symmetry of the problem that makes $V_k = -V_{N+1-k}$; because of this $V_{\frac{1}{2}(N+1)} = 0$ and there are only $\frac{1}{2}(N-1)$

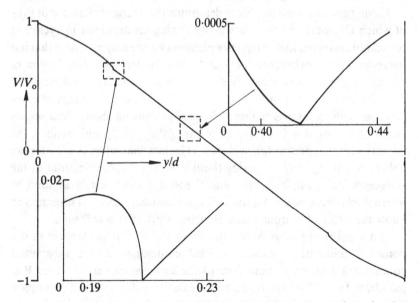

Figure 6.5 Variation of $V(y)$ when $\beta = 9.57$. The two insets show singular points greatly enlarged, after subtraction of the general linear trend.

rather than N equations. The two terms only appear simultaneously when the orbit considered straddles the centre line.

The solution of (16) when $N = 359$ and $m = 75$ is shown in fig. 5. The smooth variation of V is interrupted at the surfaces, as expected, and at one and two orbit diameters from them. In the insets the nature of the discontinuities is revealed by subtracting the mean trend of V and displaying the difference, which is very small indeed at two diameters ($y = 0.417$ and 0.583). Presumably an even fainter ghost of the surface is to be found at three diameters, but five-figure accuracy in the computation does not suffice to reveal it. The origin of the discontinuities is clear from fig. 4 and the accompanying analysis. Orbits which intersect the surface make a contribution at P quite different from those which fail, by however small a margin, to intersect. The former arrive as from a point at potential V_0, the latter as from the mean potential round the orbit. And as the weight factors w_i peak at the orbit extremes, a sharp change is to be expected as P moves through the point $y = d$. At $2d$ there is no discontinuity in orbit character, all contributing orbits bringing their mean value of V, and the discontinuity is only a pale reflection of that at d. Deep in the material V can be very well approximated by a linear function of y, for such a function satisfies (16) and only fails to describe the whole form of V because it fails to take account of the peculiarities arising from incomplete orbits.

Having gone to some trouble to determine the strange behaviour of $V(y)$, of which Chambers[26] describes many analogous examples in cyclotron resonance and other high-frequency phenomena, we may now note that it is irrelevant to the calculation of ρ_{eff}. Adding an arbitrary distribution of space-charge changes the distribution of V, but at any point only expands or contracts the Fermi surface uniformly, since the kinetic energy of every electron arriving there is changed identically. Thus no change to $\mathbf{J}$ results and we may just as well assume a constant $\partial V/\partial y$, or $\mathscr{E}_y$, and calculate the current with no concern for neutrality. The fact that when $\beta < 2$ we were able to obtain ρ_{eff} without giving thought to any but the electrons at the surfaces is an example of this general result. Unfortunately as soon as internal scattering enters this simplification vanishes, since it is necessary to know the value of V from which electrons start after a collision.

In a number of papers Azbel' and Peschanskii[27] refine (almost to the point of obliteration) Azbel's original contention that a substantial concentration of the current close to the surfaces, especially where $\mathbf{B}$ is parallel to the surface, can play a major role in conduction in thin samples. There is much truth in this idea if the surface reflects electrons specularly, for those with orbit centres lying within R of the surface can execute skipping orbits to achieve a long effective path, so that they are much more susceptible to excitation by $\mathscr{E}$ than those in closed orbits. The process is effective, however, only in a compensated metal and is eliminated by random scattering at the surface. In those unusual circumstances when some specular reflection occurs the theory given above is readily extended by modifying the boundary conditions. Thus the distribution of electrons leaving the surface may partially reproduce that of the incident electrons and for the rest be described by ε' as in (7); an example will appear later. The skipping electrons are then automatically taken into account.

Førsvoll and Holwech[28] have made careful measurements on rolled aluminium plates in this geometry, an example being shown in fig. 6 for a plate of thickness $d = 25 \ \mu$m. This was in fact the stimulus for Ditlefsen and Lothe's[25] choice of $l/d = 36$ for one of their computations, as shown in fig. 4. Direct comparison is hindered by the intrinsic magnetoresistance of

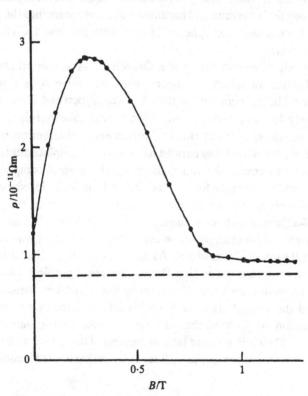

Figure 6.6 Variation with B of the resistivity of a polycrystalline aluminium plate (Førsvoll and Holwech[28]); **B** is normal to **J** and lies in the plane of the plate.

aluminium, but a liberal interpretation of Nordheim's formula (2) enables us to make a rough check, especially as it is obvious that the measured and theoretical curves are very similar. The sample purity was such that the transverse magnetoresistance should have saturated at about 0.1 T, and the broken line has been drawn to represent this saturation value, at a height that matches the theoretical high-field tail. The end-point of the peak, where $\beta = 2$, occurs at 0.80 T and the maximum at 0.29 T, which is rather larger than the theoretical value of 0.22 T. The excess resistivity at the peak is 2.0×10^{-11} Ωm, at which point the theoretical value of Q in (13) is 0.64. To make theory and experiment consistent $\rho_0 l$ must be taken as 7.8×10^{-16} Ωm^2, much the same as was needed to explain the zero-field size effect. Of course this improbably large value, as we noted before, most likely arises from small-angle scattering, whose influence on the size-effect in a magnetic field is even more conjectural than when $B = 0$. We must be satisfied, therefore, that there is general agreement without asking for more, until someone finds the time and hardihood to incorporate into the theory small-angle scattering, non-spherical Fermi surfaces, and the effects of polycrystallinity.

The original measurements by MacDonald,[19] as well as others made later, used wires for which the theory, with transverse **B**, is extremely troublesome. The ambitious attempt by Mackey, Sybert and Deering[29] is unfortunately defective on two counts. First, they assume that electrons are scattered from the surface with the equilibrium Fermi distribution; we have seen, however, that when **B** lies parallel to the surface a displacement ε' must be introduced to ensure the vanishing of J_n, the normal component of current. Secondly, they take for granted that $\mathscr{E}$ is uniform. Insofar as they neglect problems of internal charge neutrality we have seen that this is largely-self-adjusting without influence on **J**, but it is not obvious that J_n can be made to vanish at all points on the surface of a circular wire unless a more general form of $\mathscr{E}$ is assumed. As soon as one allows that $\mathscr{E}$ may be non-uniform over the cross-section, the mathematical problem moves into a different realm of complexity. The claim by Azbel' and Peschanskii[24] to have solved the general problem when l is infinite remains a claim to a formal solution only, since they do not attempt to translate it into computation. Their analysis and later extensions of the theory make heavy reading for what the experimenter will regard as rather limited recompense.

The Sondheimer and related effects

The third principal orientation of **B** to be considered is the easiest to calculate for a free-electron metal; this is the case first analysed by

Sondheimer,[30] in which **B** is normal to the plate. For a given angle relative to **B**, the electrons travel in circular helical paths just as far before striking the surface as if they had been moving in straight lines, and there is no need to introduce non-zero ε' to ensure the vanishing of J_n. The effective-path method works well here.

The theory goes just as easily for an arbitrary Fermi surface with axial symmetry about the normal to the plate, as for a free-electron sphere. Let us consider an electron on a circular section of radius k_t and thickness δk_z, and starting at a distance y from one side of the plate. Moving towards this side in a helix of radius $R = k_t/\alpha$ and pitch $Z = \alpha^{-1}(\partial \mathcal{A}_k/\partial k_z)_E$ according to (1.30), it strikes the surface after traversing y/Z turns. If its initial position in k-space is denoted by φ in fig. 7, it reaches $\varphi + 2\pi y/Z$ at the surface. On translating into real space, and using complex representation for displacements in the plane, we have that L_x and L_y are the real and imaginary parts of L, where

$$L = iRe^{i\varphi}(1 - e^{2\pi iy/Z}). \tag{6.19}$$

Since an impulsive field $\mathcal{E}_x$ creates particles at the same density for all y, we may directly average (19) over y to obtain the mean value of $L(\varphi)$:

$$\bar{L}(\varphi) = Re^{i\varphi}[i + (Z/2\pi d)(1 - e^{2\pi id/Z})]. \tag{6.20}$$

This result ignores internal collisions, which can if necessary be included without difficulty. The second term in the square brackets is oscillatory in B, since $Z^{-1} \propto B$; internal collisions damp the strength of the oscillations. The

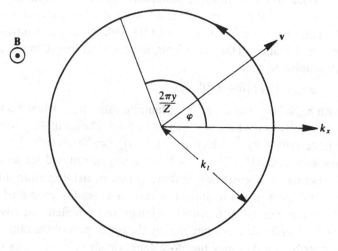

Figure 6.7 Calculation of Sondheimer effect.

corresponding conductivity follows from (1.37):

$$\delta\sigma = \delta\sigma_{xx} + i\delta\sigma_{yx} = (e^2/4\pi^3\hbar)\oint \bar{L}Re[dS],$$

$$= (e^2k_t\delta k_z/4\pi^3\hbar)\oint \bar{L}\cos\varphi \, d\varphi,$$

$$= (e\delta n/B)[i + (Z/2\pi d)(1 - e^{2\pi id/Z})], \qquad (6.21)$$

in which $\delta n = k_t^2 \, \delta k_z/4\pi^2$ and is the number of electrons per unit volume in the slice. The imaginary part, $\delta\sigma_{yx}$, oscillates about the expected mean value, $e\delta n/B$ (cf. (1.50)), while the real part oscillates between 0 and twice its mean value $\delta\bar{\sigma}_{xx} = eZ\delta n/2\pi Bd$. Since for an axially symmetric surface $Z = (2\pi k_t/\alpha)\cot\theta$, θ being the angle between $\mathbf{v}$ and $\mathbf{B}$;

$$\delta\bar{\sigma}_{xx} = (\hbar k_t\delta n/B^2 d)\cot\theta. \qquad (6.22)$$

If the slice is part of a free-electron sphere, it contributes to the high-field conductivity of an extended sample to the extent of $\hbar k_F\delta n/B^2 l$, as can be derived immediately from (1.16). Comparison with (22), k_t being written as $k_F\sin\theta$, shows that $d\sec\theta$, the distance from side to side measured along the helix, plays the part of l, a not unexpected result.

The result in (21) is of little use until it has been integrated over the whole Fermi surface. There are two cases of particular interest, the free-electron metal which was treated by Sondheimer,[30] and which has a maximum of Z at the poles; and secondly, an axially symmetric surface for which $\mathscr{A}_k(k_z)$ shows an inflexion at which Z is extremal. As Gurevich[31] pointed out, the latter should give a much more powerfully oscillatory effect.

To take Sondheimer's case first, we write $Z = (2\pi k_F/\alpha)\cos\theta$ and $\delta n = (k_F^3/4\pi^2)\sin^3\theta\delta\theta$. The integration of (21) with these substitutions, and some reorganization of the coefficients, is carried out from 0 to $\pi/2$, and the result doubled to give

$$\sigma\rho_\infty = (3/8\beta)(i + 3/8\beta + \tilde{\zeta}), \qquad (6.23)$$

in which $\rho_\infty = 3\hbar k_F/8ne^2 d$, and is the limiting value of ρ_{xx} when β is large; it is the resistivity of the bulk metal when $l = 8d/3$. The oscillatory behaviour of σ is described by $\tilde{\zeta}$, which is $-(3/2\beta)\int_0^{\pi/2}\sin^3\theta\cos\theta e^{i\beta\sec\theta} \, d\theta$, better written as $-(3/2\beta)\int_1^\infty (t^{-3} - t^{-5})e^{i\beta t} \, dt$, t being substituted for $\sec\theta$.

The evaluation of $\tilde{\sigma}$ can only be done by numerical integration, although the general form of the high-field behaviour is readily estimated. When $\beta \gg 1$, the exponential term causes the integrand to oscillate rapidly with t, and as the amplitude dies away rapidly the upper part of the range makes little contribution. We may therefore approximate $(t^{-3} - t^{-5})$ by the first term, $2(t - 1)$, of its Taylor expansion about $t = 1$, and enforce convergence by inserting a weak decrement, λ. In this limit,

$$\tilde{\sigma}/\sigma_{xx} = 8\beta\tilde{\zeta}/3 \sim -(8/\beta^2)e^{i\beta} \lim_{\lambda \to 0} \int_0^\infty z e^{(-\lambda+i)z}\,dz = (8/\beta^2)e^{i\beta}.$$

$$(6.24)$$

In a sample with internal scattering the convergence factor is automatically supplied by replacing $\beta(= d/R)$ by $\beta + id/l$. Sondheimer's theory shows that this substitution should be made throughout (23). Relative to the mean, $\bar{\sigma}_{xx}$, itself falling as $1/B^2$, the amplitude of oscillation decays as $1/B^2$. This rapid weakening of the oscillation, clearly shown by the computed forms of ρ_{xx} in fig. 8 for $d/l = 0$ and 1, is a consequence of the shape of the Fermi surface. Since Z is a monotonic function of θ the only oscillatory frequency singled out for special weight in the integral for $\tilde{\sigma}$ is that due to the limiting points, where k_z is extremal, and $\theta = 0$ and π; this is clear from the approximation (24) which depends on selecting values of t near unity. At a limiting point, however, the electron velocity has no component lying in the plane. For $\tilde{\sigma}$, therefore, one relies on electrons on the caps near the limiting points having a small component of $\mathbf{v}$ in the plane; the larger β is the thinner, and less effectively conducting, are the caps.

The frequency of the oscillations, in a graph of $\rho_{xx}(B)$, tends to a constant, in contrast to the quantum oscillations of chapter 4 which have constant frequency when plotted against B^{-1}. From (24) the period is seen to be such

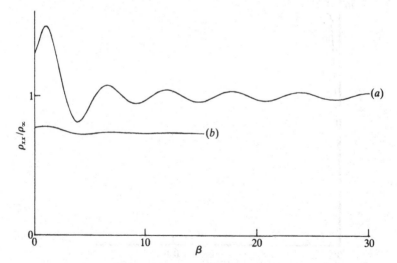

Figure 6.8 Theoretical form of Sondheimer oscillations for a free-electron metal, (a) with no internal scattering (b) with $l = d$. The vertical scale is reduced 5 times for (b).

that $\Delta\beta = 2\pi$, or $\Delta B = 2\pi\hbar k_F/ed$. More generally, (21) shows that ΔB is such that $\Delta(1/Z) = 1/d$ so that, from (1.30),

$$\Delta B = (\hbar/ed)(\partial\mathscr{A}_k/\partial k_z)_E, \qquad (6.25)$$

the derivative being taken at the limiting point. This result may be immediately extended to apply to any closed Fermi surface, which must have at least one pair of equivalent limiting points, or any unclosed surface which may happen to have limiting points. At the limiting point of a smooth surface which may be expressed as a quadratic form in the vicinity, $\partial\mathscr{A}_k/\partial k_z$ is related to the Gaussian curvature K,[32] which is the product of the two principal curvatures: $\partial\mathscr{A}_k/\partial k_z = 2\pi/K^{1/2}$ so that $\Delta B = 2\pi\hbar/edK^{1/2}$.

Excellent Sondheimer oscillations have been observed in aluminium by Førsvoll and Holwech[28] (fig. 9) who fitted their observations to the theory by adding an assumed bulk magnetoresistance. In view of the polycrystalline plates they used, and the rather sharply faceted Fermi surface of aluminium, it would be tedious to undertake a detailed discussion, and we shall be content to observe that a full analysis ought to achieve a very satisfactory correlation between theory and experiment. Earlier Babiskin

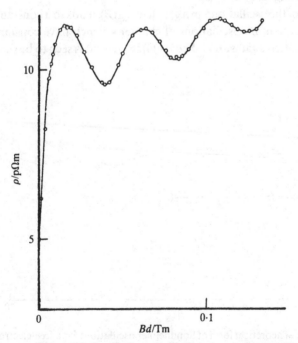

Figure 6.9 Sondheimer oscillations in a polycrystalline aluminium plate (Førsvoll and Holwech[28]).

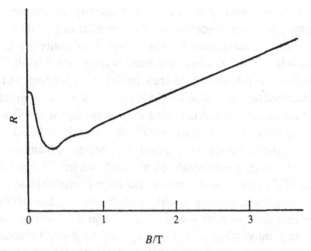

Figure 6.10 Resistance (arbitrary units) of an 80 μm sodium wire in a transverse magnetic field (Babiskin and Siebenmann[33]).

and Siebenmann[33] had seen rather less pronounced oscillations in an 80 μm circular wire of sodium (fig. 10) superimposed on the linear magnetoresistance characteristic of alkali metals. In a circular wire, of course, there is no distinction between the MacDonald and the Sondheimer geometries. It is therefore not surprising to see, with transverse **B**, the initial drop when $\beta < 2$ which is the manifestation of the former case, and the oscillations of the latter at higher values of β. Comparison of the positions of the maxima and minima with those in fig. 8 show that for the first of each the wire behaves like a plate some 12–15% thinner than the wire diameter, the difference becoming smaller for the next two or three, which is as far as the oscillations can be followed. This is very much what one would expect – only when there are many periods of the limiting-point helix within the thickness of the wire will the resonance come to be controlled by the maximum dimension. According to Babiskin and Siebenmann's quoted value for r_0 of 2000, the value of d/l for this sample was 1.14, and comparison with the curve for $d/l = 1$ in fig. 8 makes it surprising that the oscillations were so readily seen; perhaps the process of casting the wire in a capillary tube improved its purity a little. It is worth asking why MacDonald and Sarginson,[19] seven years earlier, using nearly as pure material and wires as thin as 30 μm, did not discover the oscillations. Of course, they may not have benefited from accidental enhancement of purity, but it is also possible that if they had gone (or, more likely, been able to go) to rather higher values of B the oscillations would not have escaped them.

The slow variation with B of the mean resistivity in sodium and aluminium helps the direct observations of the oscillations. If they arise from limiting points in a compensated metal the mean resistivity would rise as B^2 while the relative magnitude of the oscillations would still fall off as B^{-2}. The absolute magnitude would then be field-independent, but no more readily detected in a plot of $\rho(B)^*$. On the other hand, by modulating B with a constant amplitude and detecting the voltage signal at twice the modulation frequency, the quadratic mean variation is reduced to a constant while the oscillations, having period independent of B and a much higher second derivative, should ideally be revealed with field-independent amplitude. Fig. 11 illustrates this point well. The effect of internal scattering is to reduce the amplitude, as in fig. 8, without changing the field-variation, since after the first few cycles at low fields the integral is dominated by values of t around unity; when β in (23) is replaced by $\beta + \mathrm{i}d/l$ a sensibly constant attenuation factor, $\mathrm{e}^{-d/l}$, can be taken outside the integral. It must

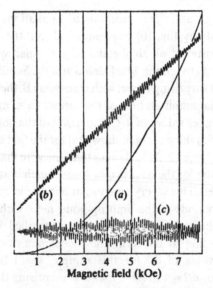

Figure 6.11 Resistance oscillations of a single-crystal gallium plate in the Sondheimer geometry; (a) R, (b) dR/dB and (c) $\mathrm{d}^2R/\mathrm{d}B^2$ (Munarin *et al.*[34]).

---

*The Hall conductivity σ_{yx} is small in a compensated metal and there is nothing to mask the imaginary component of $\bar{\sigma}$. Reynolds and colleagues[36] have concentrated on this and other transport coefficients in a single crystal plate of cadmium. They find long sequences of oscillations in σ_{yx} and some of the other coefficients.

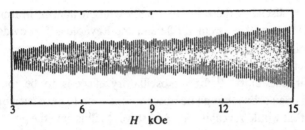

3   6   9   12   15
H kOe

Figure 6.12 Resistance oscillations of a cadmium plate in the Sondheimer geometry; d^2R/dB^2 is recorded directly (Hambourger *et al.*[35]).

be remembered, however, that at large values of β, when there are many turns of the helix in the thickness of the plate, variations in this thickness will be very effective in smearing out the oscillations, which must then be expected to decay as B increases. The samples that yielded figs. 11 and 12 were very carefully prepared with this in mind.

It is clear we must interpret the oscillations in the compensated metal Cd, shown in fig. 12, in terms other than the Sondheimer limiting-point process, which cannot yield an increasing amplitude. Here we have an example of Gurevich's process, when the helical pitch Z passes through an extremum in a region of the Fermi surface where the electrons have a substantial component of velocity in the plane of the plate. They contribute well to $\tilde{\sigma}$ and the phase of the oscillations is momentarily stationary with respect to k_z. If the Fermi surface is axially symmetrical (21) describes the conductivity due to a thin slice, the amplitude being proportional to δk_z and the phase φ of the oscillatory part being $2\pi d/Z$. At a stationary point φ varies as $\varphi_0 + a(\Delta k_z)^2$, the same relation between amplitude and phase as that we met in chapter 4, which gave rise to the Cornu spiral. If the phase at the stationary point is high enough we may take over the former argument unchanged to express the resultant oscillatory amplitude – it is as if a strip oscillated all in the same phase, the width of the strip being defined by those points where $\varphi = \varphi_0 \pm \pi/4$ (the sign depending on whether φ_0 is a maximum or a minimum). Since $d/Z \propto B$, the width varies as $B^{-1/2}$, and $\tilde{\sigma}$ from (21) as $B^{-5/2}$. In a compensated metal σ_{yx} may be taken as zero, while $\bar{\sigma}_{xx} \propto B^{-2}$. As a result $\tilde{\sigma}/\bar{\sigma}_{xx}$ and $\tilde{\rho}/\bar{\rho}_{xx}$ both vary as $B^{-1/2}$, and $\tilde{\rho}$ itself increases as $B^{3/2}$. This is a more rapid increase than is shown in fig. 12, except for a tantalizing hint at the high end that something more like theoretical expectation is about to occur.

Of course, the Fermi surfaces of gallium and cadmium are far from the axially symmetric model, but only the details are modified when realistic helices are incorporated, as also when **B** is not normal to the plate.

214

146

Chambers[26] discusses these points briefly while Munarin, Marcus and Bloomfield,[34] as well as Grenier, Efferson and Reynolds,[36] provide more detailed explorations; these last two teams give rather full references to the Western literature.

A further possibility[37] for an oscillatory effect is to be found, in principle, in open orbits though no observations seem to have been made in conditions that would reveal it. Fig. 3.21 serves to illustrate the point. Let a single crystal of copper be cut in the form of a plate lying in the plane of the diagram, i.e. with [011] normal to the plate, and let **B** lie in the plane, along [211] as shown. The periodic open orbits labelled CC' run normal to the plate and all have the same period Z, which is α^{-1} times the width CC' of the Brillouin zone. They are most strongly excited by an electric field parallel to CC' and should therefore give rise to Gurevich-like oscillations, though with **B** lying in the plane of the sample; indeed, because in this case Z is strictly constant over a slice of the Fermi surface, instead of being extremal, we expect σ/σ_{xx} to be independent of B instead of decaying as $B^{-1/2}$. Even with **J** parallel to **B** one should observe oscillations, but now the equatorial open orbit plays a part analogous to the limiting point on a closed Fermi surface; in these circumstances enhanced Sondheimer-like oscillations are to be expected. These ideas can easily be refined and extended to other types of periodic open orbit such as AA' and BB' in fig. 3.21, and to aperiodic open orbits which have no unique Z but instead a spread distribution of periodicities which might make them simulate the original Gurevich oscillations. Further analysis, based on known Fermi surfaces, may well wait until there are experiments to interpret.

Electron focussing

We discuss in this section a number of experiments which imitate in metals the focussing properties that have long been used in mass spectrography and in β-ray spectroscopy. The pioneer of their application to metals was Sharvin,[38] working in Kapitza's institute and reproducing the geometry that was suggested by Kapitza himself 45 years earlier and successfully realized by Tricker.[39] Sharvin attached electrodes to opposite sides of a single crystal plate of tin, 0.4 mm thick, having perfected the technique of welding so that the contacts were no more than 1 μm across. He passed a current into the plate from one electrode, taking it out at the edge of the sample, and measured the potential difference between the edge and the other electrode. This is shown schematically in fig. 13(a), and it will be appreciated that by this arrangement he avoided measuring the consider-

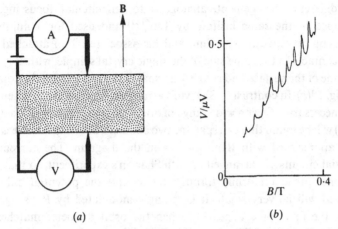

(a)   (b)

Figure 6.13 Electron focussing parallel to **B**; (a) schematic circuit diagram showing point contacts on opposite sides of the plate (shaded), (b) $V(B)$ for a plate of single-crystal tin, 0.4 mm thick, at 1.3 K with a current of 100 mA (Sharvin and Bogatina[40]).

able potential drop across the input electrode; if the current spread uniformly from the input towards the edges, the potential difference between the opposite electrode and the edge would be very small. With a magnetic field applied precisely along the line joining the electrodes an oscillatory signal was observed that indicated when the thickness of the sample equalled an integral number of periods of the helical path of the electrons. When this condition is satisfied for electrons leaving the input point at a given angle θ to **B**, they all hit the opposite electrode, irrespective of the initial azimuthal angle ϕ. Otherwise the electrons arrive at all points on a ring surrounding the potential electrode. Sharvin and Bogatina[40] observed a long sequence of oscillations, as in fig. 13(b), evenly spaced in B, and superposed on a steady trend of voltage which was insufficient to obscure the interesting signal.

To a certain extent this phenomenon resembles the Sondheimer effect, in that both depend on matching the helical period Z to the plate thickness. However, in Sharvin's experiment the potential difference is measured parallel to **B**, not normal, and this gives much greater weight to the limiting points, where Z is nearly constant over a rather wide range of θ, and where the electrons are precisely those that are most favoured for running between the electrodes. It is this that allows a long train of oscillations to be seen, much as with Gurevich's extension of the Sondheimer effect. No theory of Sharvin's oscillations seems to have been published.

212

A different arrangement, analogous to semicircular focussing, was developed in the same insitute by Tsoi,[41] and used for a number of interesting investigations by him and his associates. Tsoi attached both point contacts to the same side of the single crystal sample, with **B** lying in the plane of the surface and roughly normal to the line joining the contacts, as in fig. 14(*a*). In contrast to Sharvin's arrangement, careful alignment of **B** is not necessarily. Over a wide range of orientations, trajectories (fractional orbits) will be found that can span the two contacts, but the principle is most easily appreciated with **B** normal, as in the diagram. The current and potential circuits are arranged as with Sharvin's experiment, so that if the electrons spread uniformly through the sample the potential difference observed will be very small. It is strongly modulated by *B*, as fig. 14(*b*) shows, the first spike appearing when the orbit diameter matches the electrode separation. Then, as with Sharvin's experiment, a substantial number of electrons, having left the input electrode in a fairly broad fan around the normal to the surface, are refocussed onto the potential electrode and alter the recorded potential. In general the orbits are not semi-circular, but when the Fermi surface is closed there will be, in each section normal to **B**, a chord drawn normal to the sample surface that is extremal in length. Correspondingly in real space this chord defines the ends of a semi-orbit of extremal extension parallel to the electrodes, and it is the electrons moving in this and neighbouring semi-orbits that dominate the potential pick-up. Only one thin slice of the Fermi surface matters apreciably, being that slice on which the mean velocity along **B** serves to target the potential electrode in the *z*- as well as the *x*-direction. When **B** is normal to the line of electrodes v_z must be zero, and this is most commonly achieved when the slice is one of extremal area.

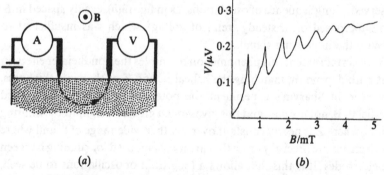

(*a*)   (*b*)

Figure 6.14 Semi-orbital focussing; (*a*) schematic circuit diagram, (*b*) *V*(*B*) for a single crystal of bismuth (Tsoi[41]).

The special interest of fig. 14(*b*) lies in the regularly spaced sequence of spikes, which arise from multiple specular reflections of electrons from the surface, the chord of the semi-orbit being a submultiple of the electrode spacing. The sample used by Tsoi to obtain this trace was a crystal of bismuth cut and electro-polished so that a principal plane lay in the surface. As already remarked, the long Bloch wavelength of conduction electrons in bismuth encourages specular reflection, and Tsoi's choice of orientation and careful surface treatment optimized the effect. Unless the surface is prepared with great care, all spikes except the first are greatly reduced or lost entirely; this accords with the most elementary view that the overall extremal property is dependent on all the component bounces of the electrons, in their journey between the electrodes being individually extremal. The total length of the extremal path is independent of *B*, and we therefore do not expect internal scattering to damp out the higher spikes preferentially, but only to smooth and diminish all equally. One may infer from the decrement of the spikes that at least 70% of electrons arriving normal to the surface are reflected specularly. 200

To test whether this intuitive picture is near enough the mark to allow the specular reflection coefficient to be quantitatively estimated, a more detailed theory is needed. In this case the formulation in two dimensions, as with fig. 1 whose notation we retain, ought to be rather reliable since only one thin slice of the Fermi surface contributes to the oscillations. We shall assume that a fraction, *p*, of electrons incident on the surface are reflected specularly, $1 - p$ being diffusively scattered with energy ε'^*. As before we ignore internal scattering, so that the theory may be formulated solely in terms of the potential distribution $V(x)$ on the surface, and the Fermi surface displacement $\varepsilon(\vartheta, x)$. The sample is assumed thick, so that every electron leaving the surface at ϑ arrives back at $-\vartheta$ and, if it is specularly reflected, leaves again at ϑ. We shall express V and ε as Fourier sums, and consider one term, of the form e^{ikx}, separately; the presence of the exponential will be taken for granted in the following analysis. In all that follows $\gamma = 2kR$, and Fourier coefficients will be distinguished by the subscript γ.

An electron striking the surface at an angle ϑ with excess kinetic energy $\varepsilon_\gamma(\vartheta)$ last left the surface a distance $2R \sin \vartheta$ to the left. Its mean kinetic energy on leaving the surface was $[p\varepsilon_\gamma(\vartheta) + (1 - p)\varepsilon'_\gamma] e^{-i\gamma \sin \vartheta}$, and it gained

---

* In a typical example of inattention to the conventions of others, *p* was introduced into the theory of the anomalous skin effect,[42] despite Fuchs'[6] and later workers' use of ε for the same quantity in treating size effects. Later *p* seems to have displaced ε everywhere, except that Tsoi prefers to use *q*.

potential energy $eV_\gamma(1 - \varepsilon^{-i\gamma\sin\vartheta})$ in traversing the arc. For compatibility,

$$[p\varepsilon_\gamma(\vartheta) + (1-p)\varepsilon'_\gamma]e^{-i\gamma\sin\theta} - eV_\gamma(1 - e^{-i\gamma\sin\theta}) = \varepsilon_\gamma(\vartheta),$$

or

$$\varepsilon_\gamma(\vartheta) = \{[(1-p)\varepsilon'_\gamma + eV_\gamma]e^{-i\gamma\sin\theta} - eV_\gamma\}/(1 - pe^{-i\gamma\sin\vartheta}). \qquad (6.26)$$

The condition for charge neutrality is that $\int_0^\pi [\varepsilon_\gamma(\vartheta) + \varepsilon_\gamma(-\vartheta)] \, d\vartheta = 0$, and already we have arrived at an uncomfortable integral to be evaluated. For our purpose great generality is not required, the most important case being that of diffuse scattering, $p = 0$, for which $\varepsilon_\gamma(-\vartheta) = \varepsilon'_\gamma$. Then this condition resolves itself into:

$$\varepsilon'_\gamma = eV_\gamma(1 - A_\gamma)/(1 + A_\gamma) \quad , \quad \text{where } A_\gamma = \pi^{-1} \int_0^\pi e^{-i\gamma\sin\vartheta} \, d\vartheta.$$

$$(6.27)$$

The evaluation of A_γ may be postponed.

The current density leaving the surface can be written immediately, the constant multiplier being supplied by the same argument as led to (11),

$$J_\gamma = (n/2\pi E_F) \int_0^\pi [\varepsilon_\gamma(-\vartheta) - \varepsilon_\gamma(\vartheta)] \sin\vartheta d\vartheta,$$

$$= (neV_\gamma/\pi E_F)(2 - \pi B_\gamma)/(1 + A_\gamma), \qquad (6.28)$$

where $B_\gamma = \pi^{-1}\int_0^\pi \sin\vartheta e^{-\gamma\sin\vartheta} d\vartheta$. To apply (28) to the experimental arrangement in which current is injected at a point, taken as the origin of x, we write ξ for $x/2R$ and

$$J(\xi) = (2\pi)^{-1} \int_{-\infty}^\infty e^{i\gamma\xi} \, d\gamma.$$

Consequently,

$$V(\xi) = (2\pi)^{-1} \int_{-\infty}^\infty (V_\gamma/J_\gamma)e^{i\gamma\xi} \, d\gamma,$$

$$= (\pi E_F/2ne)\left\{\delta(\xi) + Re\left[\int_0^\infty F(\gamma)e^{i\gamma\xi} \, d\gamma\right]\right\}, \qquad (6.29)$$

in which $F(\gamma) = (2A_\gamma + \pi B_\gamma)/\pi(2 - \pi B_\gamma)$, obtained from the reciprocal of the right-hand side of (28) by extracting as a δ-function the limiting value as $\gamma \to \infty$. This δ-function represents the steady potential associated with the current injection, which is not recorded in the arrangement used by Sharvin and Tsoi. The second term, for which we write $\tilde{V}(\xi)$, contains the measured field-dependent potential; from now on we shall pay no more attention to the absolute value of $\tilde{V}$.

The functions A_γ and B_γ are closely related to Bessel functions,

$$A_\gamma = J_0(\gamma) + iE_0(\gamma) \quad \text{and} \quad B_\gamma = E_1(\gamma) - iJ_1(\gamma), \qquad (6.30)$$

J_0 and J_1 being Bessel functions of the first kind and E_0 and E_1 Weber functions. All these are tabulated for $\gamma < 15$ in Jahnke and Emde's[43] tables, but are very quickly evaluated by numerical integration. The most important feature of $\tilde{V}$ can be seen without computation from the behaviour of $F(\gamma)$ at large values of γ. Then E_0 is approximated by the Bessel function $- Y_0$ and E_1 by $- Y_1$. As a result, at large γ, A_y tends towards the Hankel function $H_0^{(2)}(\gamma)$, which itself is asymptotically equal to $(2/\pi\gamma)^{-1/2}e^{-i(\gamma-\pi/4)}$, while B_y tends towards $-iH_1^{(2)}$, with the same asymptotic form as A_y. Since in this limit the denominator of $F(\gamma)$ tends to a constant, 2π, we have:

$$F(\gamma) \approx \text{const.} \, \gamma^{1/2} \, e^{-i(\gamma-\pi/4)}. \tag{6.31}$$

Inserting this into (29) we see that when $\xi = 1$, and only then, the integrand becomes $\gamma^{1/2}$ and the integral diverges at ∞. The theory predicts a sharp spike when the electrode separation equals the orbit diameter provided, of course, that the direction of **B** allows electrons (or holes) injected at the current input to reach the potential probe.

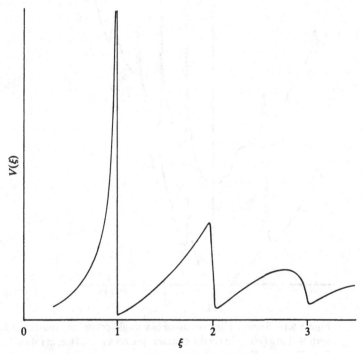

Figure 6.15 Computation of $V(\xi)$ according to (29).

To go further than this $F(\gamma)$ must be evaluated and Fourier-transformed; the result is shown in fig. 15. The first spike is as sharp as expected, with a precipitous fall on the high-field side as the orbiting electrons suddenly fail to reach the potential electrode. Subsequent spikes are progressively more rounded in form, but are stronger than might have been expected. Physically their origin lies in an effect similar to that shown in fig. 5. Where the focussed electrons strike the surface a singularity of $V(\xi)$ occurs, and the electrons leaving this point more or less normally will transfer their information of the singularity a further orbit diameter; and so on. It is not necessary to have specular reflection reflection for multiple spikes to be formed, though these are enhanced by specular reflection. The fact that Tsoi[41] does not always observe a sequence of spikes suggests that scattering into a third dimension, not to mention the spread of field lines from the singularity into the third dimension, diminishes the effect. But until the theory has been extended, at the cost of considerably heavier

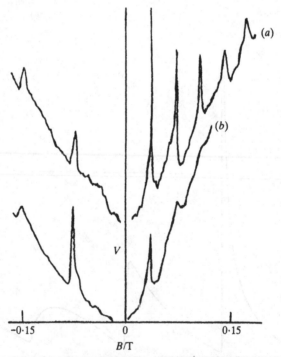

Figure 6.16 Semi-orbital focussing in a single-crystal of tungsten at 1.6 K, with **B** along [001]. In (a) the contacts are on a (110) face, in (b) on a (100) face (Tsoi and Razgunov[44]).

computation, one should be a little cautious about trusting too absolutely in the quantitative interpretation of the experiments.

All the same, the spikes seen by Tsoi and Razgunov[44] (fig. 16), using a very carefully prepared very pure tungsten (110) plane, remain sharp further out than fig. 15 suggests is compatible with $p = 0$; a significant component of specular reflection is indicated, perhaps even as much as 0.8, as they interpret the result. It may also be noted that $\pm \mathbf{B}$ give quite different patterns, the one sign of field being able to focus electrons, the other holes.

Tsoi and Razgunov note with surprise that when a tungsten plate has its surface in a (100) plane p is not more than 0.1, only the first and second spikes being discernible, and the second very weak. Since their work it has been found[45] that the (100) surface of tungsten reconstructs itself, by lateral displacement of the atoms, into a new arrangement with the unit cell twice as big in linear dimensions. This allows even a perfectly ordered surface to diffract incident electrons into sidelobes, so that there is reason to expect that the low value of p may not simply be explained away, but may turn out to be a tool to probe a most interesting phenomenon in surface physics.

After this it is something of an anticlimax to draw attention to another related study of semi-orbital focussing, which uses a pair of intersecting combs instead of two point contacts. In the original proposal[46] and the later experimental test[47] the resistance between the combs was the quantity studied, without the benefits of the Sharvin and Tsoi arrangement to eliminate the contact resistance and discriminate between electrons and holes. The observed oscillations are puny in comparison with the Russian work, and need not be discussed further. There is no reason why a different electrode arrangement should not be adopted to get round the problem of contact resistance, or field-modulation be applied as for figs. 11 and 12, but Tsoi's extraordinary successes suggest the effort would not reward time spent on alternative schemes.

Quantum effects in very small wires and rings

The effects to be discussed in this last section have only recently been discovered and are still under active investigation. Any attempt at a full treatment of the theory at this stage would be very involved and almost certainly incomplete. There are, however, some general principles that are easily appreciated and we shall deal with these alone. A review, with over 190 references, has been given by Imry[48] who has been closely associated with much of the theoretical work.

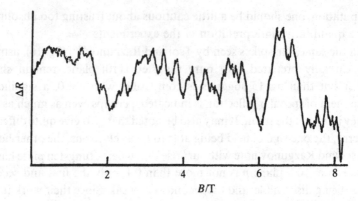

Figure 6.17 Field-variation of resistance of a very fine wire of $Au_{60}Pd_{40}$ at 0.11 K (Umbach *et al.*[49]); the vertical range of ΔR is about 2×10^{-3} of the mean resistance.

Fig. 17 shows the effect of a transverse magnetic field on the resistance, R, of an extremely thin wire of the resistive alloy $Au_{60}Pd_{40}$, formed by electron-beam lithography. The cross-section was about $60\,nm \times 38\,nm$ and the length $790\,nm$. The aperiodic fluctuations of the resistance from its mean, $R_0 \sim 100\,\Omega$, have an RMS value of about $2 \times 10^{-4}R_0$. This particular curve was taken with the sample at a temperature of 0.11 K; warming did not change R_0 or the general pattern significantly, but the amplitude of the fluctuations fell roughly as $T^{-0.7}$.

It is helpful to express the dimensions of the sample in terms of the mean free path in the bulk material, estimated to be 2.5 nm. The length was therefore about $300\,l$ and the cross-section $24l \times 15l$. Also l is about 5 times the Bloch wavelength of the electrons, and there is no reason to expect marked localization effects such as have been extensively studied in disordered semiconductors.[50] On the other hand it is not good enough to treat the electrons as classical particles being scattered from impurities, with each scattering process independent of the others. This is because at the lowest temperatures the scattering is almost entirely elastic, without phonon processes to impose random phase changes on the electronic wavefunction. A Bloch wave emitted from a point P_1 and suffering multiple scattering from points P_2, P_3, etc. before arriving at P_n is very likely to arrive with its phase perfectly determined by the length of the polygonal path joining the scattering centres p_1-p_n and by the phase change at each scattering. Another path $P_1Q_2Q_3 \ldots P_n$, starting and finishing at the same points as the first, generates a second wave at P_n that is phase-coherent with the first; and the resultant wave scattered by P_n will have an amplitude that is the vector sum of the two, just as in a Young's interference experiment. The precise disposition of the scattering centres is then crucial to the magnitude of the result – it is enough to move one by a fraction of a wavelength to make a significant change.

In fact there are not simply two paths connecting the first scattering centre with the last. On the contrary, there is an infinite number of paths available for an electron on its way through the wire, for nothing prevents the scattering sequence from returning time and time again to the same point. The greater the number of scattering centres involved, the weaker the emergent wavelet and the longer the time taken in the passage, so that residual inelastic processes may ultimately destroy phase coherence. But the wave description of an electron entering the wire, being first scattered at P_1 and then at many other points, and before suffering its last scattering at a point P_n, involves at very low temperatures the coherent superposition of an enormous number of extremely weak wavelets, each of which is the outcome of a complicated path that is very many wavelengths long. And the amplitude of the resultant is given by the vector sum of all these contributions. Since on average all phases are equally likely to be represented among the different paths, the well-known result[51] may be applied without further thought, that the resultant intensity (measured by the square of the amplitude) has a mean value given by summing the individual intensities.

This result, however, only describes the expected mean without indicating the range of fluctuations from it that may occur in specific cases. If we subject all the wavelets to random phase variations, the resultant intensity exhibits a distribution whose form is independent of the number of contributing wavelets, provided this number is reasonably large. Writing $\mathscr{P}(p)\,\mathrm{d}p$ for the probability that the emergent intensity at P_n is related to the incident intensity at P_n by a factor in the range p to $p + \mathrm{d}p$, we have that

$$\mathscr{P}(p)\mathrm{d}p = (2\pi\bar{p}p)^{-1/2}\mathrm{e}^{-p/2\bar{p}}\mathrm{d}p, \tag{6.32}$$

in which $\bar{p}$ is the mean intensity. It follows that the mean square value of the fluctuations of p about $\bar{p}$ is given by

$$\sigma^2(p) \sim \langle (p - \bar{p})^2 \rangle = \overline{p^2} - \bar{p}^2 = 2\bar{p}^2, \tag{6.33}$$

and the standard deviation $\sigma(p)$ (not to be confused with conductivity) is $\sqrt{2}\bar{p}$. Fluctuations of intensity comparable to the mean intensity itself are the origin of the analogous phenomenon of the speckle pattern observed when coherent laser light is scattered from a rough surface.[52] In this case different phase relationships are found in different regions of the pattern,

230   *Size effects*

while with the conduction problem only one possibility from the ensemble of relationships is realized – so long as $B = 0$.

We are thus led to consider the influence of B on the phase relationships, which we shall tackle before reverting to the obvious question arising from (33) – why are the observed fluctuations of resistance so much smaller than those predicted by this analysis?

The treatment of quantum effects in chapter 4 is relevant here. It was shown that the phase length of a given path is dependent on the gauge chosen for the vector potential **A**. No measurable effect results from different choices, and we may therefore make our choice for convenience. In considering a wave spreading from a scattering centre P_1 to a second centre P_2 we choose a circulating gauge, as defined by (4.8), with P_1 as gauge centre. This changes the wavenumber **k** from $mv/\hbar$ to $mv/\hbar + e\mathbf{A}/\hbar$, and since **A** is normal to the line P_1P_2 the phase length between the two is unaffected. We now consider the path P_2P_3, switching the gauge centre to P_2 for this purpose. At this point we must insert a phase correction ε, as given by (4.13) with the line P_1P_2 defining the vector **S**:

$$\varepsilon_{AB} = -\tfrac{1}{2}(\alpha \wedge \mathbf{S}_{12})\cdot\mathbf{r} = \tfrac{1}{2}\alpha\cdot(\mathbf{r} \wedge \mathbf{S}_{12}). \tag{6.34}$$

If we project the path onto a plane normal to **B**, and interpret $\mathbf{S}_{12}$ as the projected length, $\tfrac{1}{2}\mathbf{r} \wedge \mathbf{S}_{12}$ is the area of the triangle formed by O, P_1 and P_2, the origin O being arbitrarily chosen. Thus the phase correction to the whole path $P_1 - P_n$ is α times the area defined by the polygon $P_1P_2P_3...P_n$. A different path between P_1 and P_n suffers a different correction, and the two paths are phase-corrected by an amount which differs by α times the polygonal area enclosed between them. Alternatively stated,

$$\Delta\varepsilon = e\Phi/\hbar, \tag{6.35}$$

where Φ is the flux contained between the two paths. In this form the need to project the paths onto a plane is avoided. This result lies at the heart of the Aharonov–Bohm[53] effect and the quantization of flux in orbits (as discussed in chapter 4) and in superconducting rings.[54] The Aharonov–Bohm effect seems peculiarly paradoxical in that it is possible to shift the phases without apparently informing the electron of the presence of **B**. But the paradox lies deeper than this, in the fact that in all these quantum phenomena it is the unobservable (indeed, incompletely defined) **A** and not **B** that controls events.

It is now clear that when **B** is changed the phase relationships of all the paths between two points P_1 and P_n are also changed, so that the resultant wave emerging from P_n samples a fair selection of the distribution (32). The magnitude of **B** needed to affect the resultant is easily estimated as something like that which will change the flux contained in the sample by one flux quantum $2\pi\hbar/e$, or $4.14 \times 10^{-15}\,\text{Tm}^2$, since this will introduce a phase difference of 2π between paths that run close to opposite sides. For the sample of fig. 17 the area was $790 \times 60\,(\text{nm})^2$, about $5 \times 10^{-14}\,\text{m}^2$, and one expects changes of **B** amounting to 0.1 T to determine the characteristic scale of the fluctuations. This appears to be a slight underestimate but it is, after all, only an intuitive guess; Landauer and Buttiker[55] conclude tentatively, after a much more serious analysis, that something more like 2 flux quanta in the sample are needed to make a significant change. A precise statement would specify the autocorrelation function[56] $\Gamma(\Delta B)$ of the resistance at two field strengths differing by ΔB, but this has been neither determined experimentally nor calculated theoretically.

Estimating the magnitude of the resistance fluctuations is a still more difficult problem. The following argument shows clearly why it must be much smaller than (33) suggests. It was assumed in deriving (33) that the electrons entered the sample, for example from a resistanceless section, at a single point P_1 and similarly left after a final scattering by a single point P_n. In reality there are many scattering centres near the output end of the wire which could have been chosen as P_n, and each will radiate a wavelet that is more or less uncorrelated in phase and in fluctuations of phase with the others. Accordingly the fluctuations in the resultant of all outputs will be less than those of P_n alone by a factor of about $\sqrt{N}$, if there are N independent outputs. The same argument applies to the input end, to give a further factor $\sqrt{N}$, so that instead of the relative magnitude of the fluctuations amounting to $\sqrt{2}$, as in (33), we might expect $\sqrt{2/N}$. And since the conductance $G(\equiv 1/R)$ of the wire is proportional to the probability that an electron entering at one end will emerge at the other, we suppose the fluctuations in G to be given by:

$$\sigma(G) \sim \sqrt{2\,G/N}. \tag{6.36}$$

The problem is to estimate N; that is, the effective number of independent wavelets that contribute to the emergent wave. It is tempting to equate this to the number of scattering centres (Pd atoms, say) within a free path of the end of the wire, but if these are closer together than one wavelength they cannot be considered independent. Perhaps the best one can do without a radical attack on the whole problem from a more fundamental point of view,[55] is to ask how many Fourier components are needed to define completely the amplitude and phase distribution of the emergent wave; presumably this gives an upper limit to the number of sources that can be supposed to act independently. Moreover this number is easily calculated as π/λ^2 times the cross-sectional area, λ being the Bloch wavelength of the electrons, about 0.5 nm. On this assumption $N \sim 28\,000$ for the sample of fig. 17, so that $\sigma(G) \sim G/20\,000$, which is some 5–10 times less than the observed fluctuations. One should attach little importance to such a discrepancy between an exceedingly sketchy theory and a single experimental result on a sample whose very construction is something of a technical triumph.

Before leaving this aspect of the topic it is worth commenting on the temperature variation of the fluctuations, for which the quoted form $T^{-0.7}$ should surely not be taken too seriously. Up to this point we have assumed all electrons involved in the conduction process to have precisely the same wavelength, but as the Fermi tail is broadened with rising temperature this assumption must be queried. Variations of λ are unimportant provided they do not seriously affect the phase length of the meandering path of an electron through the wire. A typical path length can be estimated by use of the standard result for a random walk, that to travel a resultant distance L by means of steps each of length l the distance measured along the path must be about L^2/l. If we interpret l here as the mean free path, being that distance in which an electron loses memory of its point of origin, we find in the present case a typical path length of $5 \times 10^5 \lambda$. Taking the Fermi energy of the electrons as equivalent to a temperature of 6×10^4 K, we infer that the Fermi tail at 0.12 K is just wide enough to induce some small degree of phase-smearing. Thus we expect the curve in fig. 17 to represent fairly well the limiting form as $T \to 0$. At higher temperatures we may divide the tail into bands of such width, say 0.1 K, that each can be treated as monoenergetic, but giving rise to independent fluctuations of G. The number of bands is proportional to T and the resultant fluctuations of G are reduced by a factor $T^{-0.5}$, which seems to be taken by several authors[57] as a likely form of the general law. It is somewhat disturbing to see how the major features of the curve for 0.11 K persist, diminished only to about a quarter, up to 1.9 K at which temperature the band that generated them is presumably no more than one out of 10–20 independent bands. This fact in itself makes one doubt whether these larger features ought to be treated as part of the simple fluctuation pattern; the caveat at the end of the last paragraph deserves to be repeated.

Having spent some time on the random fluctuations we may deal rather more briefly with the effects which have generated greater interest, and which are exemplified by fig. 18. The gold ring, of mean diameter 784 nm, is

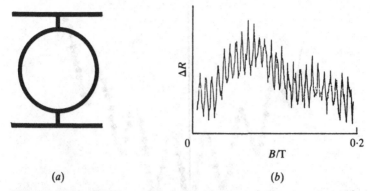

(a)   (b)

Figure 6.18 (a) Fine-line gold ring with 4-terminal arrangement for resistance measurement; (b) oscillations of R with **B** normal to the plane of the ring, at 0.01 K (Webb *et al.*[58]). The amplitude is less than 0.1% of the mean resistance. Each short period corresponds to a change of $2\pi\hbar/e$ in the flux through the ring.

made of wire whose cross-section is much the same as that used for fig. 17. It is easier to explain the period of the oscillations, which corresponds to increments of flux through the ring amounting to $2\pi\hbar/e$, than it is to carry out the experimental demonstration. Each of the two semicircular arms joining the electrodes behaves like the wire just discussed, in that there is a considerable measure of phase coherence between the input and output wave-functions. The two outputs therefore interfere constructively or destructively according to the phase difference. In fact, if the two arms were arranged to lie beside each other they would simply form part of a single thicker wire. While application of **B** normal to the ring gives rise to a relatively slowly varying output from each wire, the large contained area between the two causes their phase difference to oscillate more rapidly; the observed period was 7.6 mT as against the 200 mT characteristic of the aperiodic fluctuations which form the underlying waviness in the figure.

By contrast to fig. 18, which represents an extension of the ideas used to explain the single-strip results, fig. 19 shows oscillations of resistance at twice the frequency, the period being determined by a half-quantum of flux, $\pi\hbar/e$. There is no reason why the $2\pi\hbar/e$ oscillations should not be observed in the same ring, but we shall not go into the question of the best conditions for observing each. The second effect was predicted by Al'tshuler, Aronov and Spivak,[60] and first demonstrated by Sharvin and Sharvin.[61] It has its origin in the change of scattering cross-section that results from coherent interchange of waves between scattering centres. The simplest possible example will serve to indicate what is involved. Let us imagine two centres,

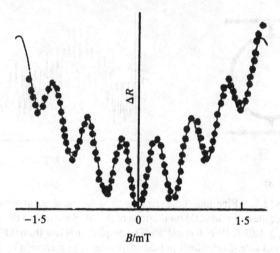

Figure 6.19 Resistance oscillations in an aluminium ring similar to that in fig. 18(a) but nearly 3 times larger (Chandrasekhar et al.[59]). Each period corresponds to a flux change of $\pi h/e$.

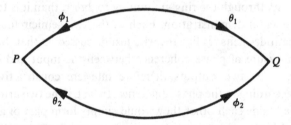

Figure 6.20 Phase lengths of different paths between P and Q.

P and Q, between which waves may travel coherently by two different paths, as in fig. 20. Thus a wavelet of unit amplitude, setting out from P along the first path, excites a secondary wavelet whose amplitude and phase are given by $r_1 e^{i\theta_1}$. The reverse process, where unit wavelet from Q travels back along the first path to P, excites there a secondary wavelet $r_1 e^{i\phi_1}$. When **B** is present θ_1 and ϕ_1 need not be equal. The corresponding phase changes for the second path are θ_2 (from Q to P) and ϕ_2 (from P to Q), and the amplitude is r_2.

When an incident wave excites unit wavelet at P the multiple scattering processes add up to Q emitting a wavelet of complex amplitude a_Q, while the initial wavelet from P is supplemented by a_P, generated by a_Q travelling along the two paths. The conditions for compatibility of these amplitudes follow immediately:

$$a_Q = (1 + a_P)(r_1 e^{i\theta_1} + r_2 e^{i\phi_2})$$

and

$$a_P = a_Q(r_1 e^{i\phi_1} + r_2 e^{i\theta_2}).$$

$$\left. \right\} \qquad (6.37)$$

Solving for a_P, we have for the total amplitude of the wavelet scattered by P,

$$1 + a_P = [1 - (r_1 e^{i\theta_1} + r_2 e^{i\phi_2})(r_1 e^{i\phi_1} + r_2 e^{i\theta_2})]^{-1}. \qquad (6.38)$$

Now let $\theta_1 = \chi_1 + \varepsilon_1$, $\phi_1 = \chi_1 - \varepsilon_1$, and similarly for θ_2 and ϕ_2. The scattered intensity then takes the form

$$|1 + a_P|^2 = |1 - r_1^2 e^{2i\chi_1} - r_2^2 e^{2i\chi_2} - 2r_1 r_2 e^{i(\chi_1 + \chi_2)} \cos(\varepsilon_1 + \varepsilon_2)|^{-2}. \qquad (6.39)$$

In this expression χ_1 and χ_2 are the mean phase lengths of the paths, which may be very large and are equally likely to take any value *modulo* 2π; $\varepsilon_1 + \varepsilon_2$, on the other hand, is half the phase difference between clockwise and anticlockwise path, or $e\Phi/\hbar$ as in (35). To find the mean value of $|1 + a_P|^2$ for different choices of P and Q, but the same Φ, we average (39) over χ_1 and χ_2.

$$\langle |1 + a_P|^2 \rangle = [1 + r_1^4 + r_2^4 + 4r_1^2 r_2^2 \cos^2(e\Phi/\hbar)]^{-1}. \qquad (6.40)$$

The mean scattering is least when Φ is an integral multiple of $\pi\hbar/e$, and greatest when it is a half-integral multiple, so that the period $\Delta\Phi$ is $\pi\hbar/e$.

The mechanism for the oscillations in fig. 19 can now be appreciated. By making the sample as a ring of thin wire a large number of paths are created that enclose very nearly the same area, and modify the mean scattering power of each centre in the same way. Thus the resistance oscillates with B, the period corresponding to a half-integral step in the number of flux quanta contained in the ring. A wavelet scattered at a certain point may find its way round the ring in either direction until it returns to its starting point. What matters on the average is not that it shall return with the same phase as it started, but that the waves in the two senses shall return in phase with one another, irrespective of what this phase may be relative to the initial phase. The sign of B is irrelevant, in contrast to the aperiodic fluctuations in fig. 17 which are quite different in form when B is reversed. There is indeed no reason to expect the behaviour when $B = 0$ to be in any way special where the aperiodic effect is concerned. On the other hand (40) implies symmetry with respect to $\pm B$, with a minimum in the resistance when $B = 0$; both implications are realized in fig. 19. As in the aperiodic effect, the oscillations in fig. 18 need not be symmetrical in B, since when $B = 0$ the resultant phase difference, for the multiplicity of paths between

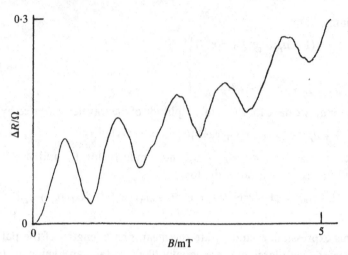

Figure 6.21 Resistance oscillations of a thin-walled magnesium cylinder in a longitudinal field at 1.12 K (Sharvin and Sharvin[61]). The tube was 1 cm long and had diameter 1.5 μm, with a mean resistance of 15 kΩ. Each period corresponds to a flux change of $\pi\hbar/e$.

the electrodes round the two sides of the ring, is not constrained to any particular value.

The same explanation accounts for the observations of Sharvin and Sharvin.[61] They evaporated a thin film of magnesium onto a fine quartz filament, about 1.5 μm in diameter and 1 cm long, and studied how the resistance in the lengthwise direction varied with **B**, applied parallel to the filament. The variations in a typical sample, with amplitude only 4 parts per million of the mean resistance, are shown in fig. 21. The period corresponds to the half-quantum of flux, $\pi\hbar/e$, as expected from (40).

At this point, after an exposition of recently discovered effects, tiny in magnitude but interesting in substance and still only imperfectly understood, we may leave the general field of magnetoresistance in which much has been accomplished and much left incomplete. But there is no lack of hints for future development, if there are any inspired to reopen the subject.

References

Preface

1. L.L. Campbell, Galvanomagnetic and thermomagnetic effects (London, Longmans, Green, 1923).

Chapter 1

1. A. Leduc, *Compt. Rend.* **102**, 358 (1886).
   B. Donovan and G.K.T. Conn, *Phil. Mag.* **40**, 283 (1949).
2. P.L. Kapitza, *Proc. Roy. Soc.* A**119**, 358 (1928).
3. P.L. Kapitza, *Proc. Roy. Soc.* A**123**, 292 (1929).
4. J. de Launay, R.L. Dolecek and R.T. Webber, *J. Phys. Chem. Solids* **11**, 37 (1959).
5. F.R. Fickett, *Ann. Report, Project 186*, International Copper Research Association (1972).
6. P.B. Alers and R.T. Webber, *Phys. Rev.* **91**, 1060 (1953).
7. D. Shoenberg, *Magnetic oscillations in metals*, p. 228 (Cambridge University Press, 1984).
8. L. Taillefer, R. Newbury, G.G. Lonzarich and J.L. Smith, *J. Magn. Magn. Mat.* **63 & 64**, 372 (1987).
9. E.H. Hall, *Am. J. Math.* **2**, 287 (1879).
10. A.M. Simpson, *J. Phys.* F**3**, 1471 (1973).
11. J.L. Olsen, *Electron transport in metals*, p. 67 (New York, Interscience, 1962).
12. N.E. Alekseevskii and Yu.P. Gaidukov, *JETP* **36**, 311 (1959).
13. H.H. Potter, *Proc. Roy. Soc.* A**132**, 560 (1931).
14. D.K.C. MacDonald and K. Sarginson, *Proc. Roy. Soc.* A**203**, 223 (1950).
15. J.A. Munarin and J.A. Marcus, *Low Temperature Physics (LT9)* p. 743 (New York, Plenum, 1965).
16. R.W. Stark, *Phys. Rev.* **135**, A1698 (1964).
17. W.B. Willott, *Phil. Mag.* **16**, 691 (1967).
18. See ref. 7, p. 11.
19. See ref. 7, p. 22 and *passim*.
20. A.B. Pippard, in *Electrons in metals*, p. 2 (ed. J.F. Cochran and R.R. Haering; New York, Gordon and Breach, 1968).
21. W. Shockley, *Phys. Rev.* **79**, 191 (1950).
22. L. Onsager, *Phil. Mag.* **43**, 1006 (1952).
23. R.G. Chambers, *Proc. Roy. Soc.* A**215**, 481 (1952).

238   *References*

24. R.G. Chambers, in *The physics of metals; 1, electrons*, p. 176 (ed. J.M. Ziman, Cambridge University Press, 1969); and in Electrons at the Fermi Surface, p. 102 (ed. M. Springford; Cambridge University Press, 1980).
25. M. Kohler, *Ann. Phys.* **32**, 211 (1938).
26. J.C. Garland and R. Bowers, *Phys. Rev.* **188**, 1121 (1969).
27. A.H. Wilson, *The theory of metals*, p. 226 (Cambridge University Press, 2nd edition 1953).
28. L. Onsager, *Phys. Rev.* **37**, 405 (1931) and **38**, 2265 (1931).
29. e.g. J.M. Ziman, *Electrons and phonons*, p. 512 (Oxford, Clarendon Press, 1960) and A.B. Pippard, *Rep. Prog. Phys.* **23**, 176 (1960).
30. A.N. Gerritsen, W.J. de Haas and P. van der Star, *Physica* **9**, 241 (1942).
31. C.M. Hurd, *The Hall effect in metals and alloys*, p. 232 (New York, Plenum, 1972).

Chapter 2

1. R.V. Jones and C.W. McCombie, *Phil Trans. Roy Soc.* *A***244**, 205 (1952).
2. J.R. Klauder, W.A. Reed, G.F. Brennert and J.E. Kunzler, *Phys. Rev.* **141**, 592 (1966). Also ref. 1.31, p. 183.
3. O.V. Lounasmaa, *Experimental principles and methods below 1 K*, p. 140. (London, Academic Press, 1974).
4. J.A. Delaney and A.B. Pippard, *Rep. Prog. Phys.* **35**, 677 (1972).
5. Y. Ueda and T. Kino, *J. Phys. Soc. Japan* **48**, 1601 (1980).
6. J.S. Lass, *J. Phys.* *C***3**, 1926 (1970).
7. R. Fletcher, *Phys. Rev. Lett.* **45**, 287 (1980).
8. G.J.C.L. Bruls, J. Bass, A.P. van Gelder, H. van Kempen and P. Wyder, *Phys. Rev. Lett.* **46**, 553 (1981).
9. P.M. Gerhart and R.J. Grass, *Fundamentals of fluid mechanics*, p.610 (Reading, Mass., Addison-Wesley, 1985).
10. J.H. Jeans, *Electricity and magnetism*, p. 266 (Cambridge University Press, 5th edition 1925).
11. R.F. Wick. *J. App. Phys.* **25**, 741 (1954).
    H.J. Lippmann and F. Kuhrt, *Z. Naturforsch.* **13a**, 462 (1958).
    W. Schneider, H. Bruhns and K. Hübner, *J. Phys. Chem. Solids* **41**, 313 (1980).
12. W.A. Reed, E.I. Blount, J.A. Marcus and A.J. Arko, *J. App. Phys.* **42**, 5453 (1971).
13. C. Herring, *J. App. Phys.* **31**, 1939 (1960).
14. J.S. Lass, Ph.D. dissertation, University of Cambridge (1969).
15. D. Stroud and F.P. Pan, *Phys. Rev.* *B***13**, 1434 (1976) and **20**, 455 (1979).
16. J.A. Stratton, *Electromagnetic theory*, p. 211 (New York, McGraw-Hill, 1941).
17. H. Stachowiak, *Physica* **45**, 481 (1970).
18. Ref. 16, p. 216.
19. P.M. Martin, J.B. Sampsell and J.C. Garland, *Phys. Rev.* *B***15**, 5598 (1977).
20. J.M. Ziman (ref. 1.29) p. 393; and ref. 1.17.
21. S.G. Lipson, *Proc. Roy. Soc.* *A***293**, 275 (1966).
22. M.A. Archibold, J.E. Dunick and M.H. Jericho, *Phys. Rev.* **153**, 786 (1967).
    M.R. Stinson, R. Fletcher and C.R. Leavens, *Phys. Rev.* *B***20**, 3970 (1979).

23. C.P. Bean, R.W. De Blois and L.B. Nesbitt, *J. App. Phys.* **30**, 1976 (1959).
24. L.D. Landau and E.M. Lifshitz, *Electrodynamics of continuous media*, p. 193 (Oxford, Pergamon, 1960).
25. M.I. Grossbard, Ph.D. dissertation, University of Cambridge (1978).
26. Ref. 1.10.
27. G.W. Ford and S.A. Werner, *Phys. Rev.* **B8**, 3702 (1972).
    S.A. Werner and G.W. Ford, *Phys. Rev.* **B11**, 1772 (1975).
28. V.T. Petrashov, *Rep. Prog. Phys.* **47**, 47 (1984).
29. K.G. Budden, *Radio waves in the ionosphere*, p. 47 (Cambridge University Press, 1961).
30. R.G. Chambers and B.K. Jones, *Proc. Roy. Soc.* **A270**, 417 (1962).
31. C. Legéndy, *Phys. Rev.* **A135**, 1713 (1964).
32. J.P. Klozenberg, B. McNamara and P.C. Thonemann, *J. Fluid Mech.* **21**, 545 (1965).
33. D.E. Chimenti and B.W. Maxfield, *Phys. Rev.* **B7**, 3501 (1973).
34. M.J. O'Shea and M. Springford, *Phys. Rev. Lett.* **46**, 1303 (1981).
35. M. de Podesta and M. Springford, *J. Phys.* **F17**, 639 (1987).
36. J.A. Delaney and A.B. Pippard, *J. Phys.* **C4**, 435 (1971).
37. M.I. Grossbard, *J. Phys.* **F9**, 1833 (1979).
38. P.B. Visscher and L.M. Falicov, *Phys. Rev.* **B2**, 1518 (1970).
39. For early references, see F.W. Holroyd and W.R. Datars, *Can. J. Phys.* **53**, 2517 (1975).
40. C. Verge, Z. Altounian and W.R. Datars, *J. Phys.* **E10**, 16 (1977).
41. A.E. Dixon, *Phys. Rev.* **B12**, 1200 (1975).
42. J.S. Lass and A.B. Pippard, *J. Phys.* **E3**, 137 (1970).
43. J.S. Lass, *Phys. Rev.* **B13**, 2247 (1976).
44. P.B. Visscher and L.M. Falicov, *Phys. Rev.* **B2**, 1522 (1970).

Chapter 3

1. For a general review see F.J. Blatt, *Solid State Physics*, **4**, 199 (ed. F. Seitz and D. Turnball, New York, Academic Press, 1957).
2. F. Seitz, *Phys. Rev.* **79**, 372 (1950).
3. Ref. 1.31, p. 153.
4. Ref. 1.21.
5. R. Olson and S. Rodriguez, *Phys. Rev.* **108**, 1212 (1957).
6. A.B. Pippard, *Phil. Trans. Roy. Soc.* **A250**, 325 (1957).
7. See ref. 1.7, p. 196.
8. M. Tsuji, *J. Phys. Soc. Japan* **13**, 979 (1958).
9. See C.E. Weatherburn, *Differential geometry of three dimensions*, pp. 69 and 226 for a formal discussion (Cambridge University Press, reprinted 1964).
10. H. Jones and C. Zener, *Proc. Roy. Soc.* **A145**, 268 (1934).
11. Ref. 1.27 and 1.29.
12. One text that avoids this criticism is N.W. Ashcroft and N.D. Mermin, *Solid State Physics* (New York, Holt Rinehart and Winston, 1976). See p. 131 for the independent particle picture, and p. 329 for a critique and justification.
13. W.A. Harrison, *Phys. Rev.* **116**, 555 (1959).
14. E. Fawcett, *Adv. Phys.* **13**, 139 (1964).

240   *References*

15. A.P. Cracknell and K.C. Wong, *The Fermi surface* (Oxford, Clarendon, 1973).
16. Ref. 1.7.
17. N.W. Ashcroft, *Phil. Mag.* **8**, 2055 (1963).
18. V. Heine in *The physics of metals* (ref. 1.24) p. 1.
19. G.G. Lonzarich and P.M. Holtham, *Proc. Roy. Soc.* A**400**, 145 (1985).
20. A.V. Gold, *Phil. Trans. Roy Soc.* A**251**, 85 (1958).
21. I.M. Lifshitz and V.G. Peschanskii, *J.E.T.P.* **35**, 1251 (1958).
22. Ref. 2.2.
23. I.M. Lifshitz, M.Ya. Azbel' and M.I. Kaganov, *J.E.T.P.* **3**, 143 (1956) and **4**, 41 (1957).
24. B. Lengeler and C. Papastaikoudis, *Phys. Rev.* B**21**, 4368 (1980).
25. Ref. 2.21.
26. R.G. Chambers in *The Fermi surface*, p. 113 (ed. W.A. Harrison and M.B. Webb; New York, Wiley, 1960).
27. J.S. Moss and W.R. Datars, *Phys. Lett.* **24A**, 630 (1967).
28. R.G. Chambers, *Proc. Roy. Soc.* A**238**, 344 (1956).
29. M. Kohler, *Ann. Phys. Lpz.* **5**, 99 (1949).
30. A.B. Pippard, *Phil. Trans. Roy. Soc.* A**291**, 569 (1979).
31. R.L. Powell, Ph.D. dissertation, University of Cambridge (1966).
    R.L. Powell, A.F. Clark and F.R. Fickett, *Phys. Kond. Mat.* **9**, 104 (1969).
32. J.O. Ström-Olsen, *Proc. Roy. Soc.* A**302**, 83 (1967).
33. A.B. Pippard, *Proc. Roy. Soc.* A**282**, 464 (1964).
34. P.G. Klemens and J.L. Jackson, *Physica* **30**, 2031 (1964).
35. F.R. Fickett, *Phys. Rev.* B**3**, 1941 (1971).
36. A.B. Pippard, *Proc. Roy. Soc.* A**305**, 291 (1968).
    R.A. Young, J. Ruvalds and L.M. Falicov, *Phys. Rev.* **178**, 1043 (1969).
37. J.C. Garland and R. Bowers, *Phys. Rev.* **188**, 1121 (1969).
38. E.E. Pacher, *Phys. Stat. Sol.* B**64**, K29 (1974).

Chapter 4

1. L.W. Schubnikov and W.J. de Haas, *Proc. Netherlands Royal Acad. Sci.* **33**, 130 and 163 (1930).
2. M.H. Cohen and L.M. Falicov, *Phys. Rev. Lett.* **7**, 231 (1961).
3. J.W. Leech, *Classical mechanics*, pp. 20 and 119 (2nd edition, London, Science Paperbacks, 1965).
4. W. Kohn, *Phys. Rev.* **115**, 1460 (1959). For other references, see L.M. Roth, *Phys. Rev.* **145**, 434 (1966).
5. Ref. 1.22.
6. Ref. 1.7, chapter 1.
7. L.M. Roth, Ref. 4. See also R.G. Chambers, *Proc. Phys. Soc.* **89**, 695 (1966) and H.H. Hosack and P.L. Taylor, *Phys. Rev.* B**3**, 4091 (1971).
8. R.B. Dingle, *Proc. Roy. Soc.* A**211**, 500 (1952).
9. Ref. 1.19.
10. J.H. van Vleck, *The theory of electric and magnetic susceptibilities*, p. 100 (Oxford University Press, 1932).
11. Ref. 1.7, p. 65.

12. Ref. 1.7, p. 41.
13. S.G. Lipson and H. Lipson, *Optical physics*, p. 154 (Cambridge University Press, 1969).
14. Ref. 1.7, p. 153.
15. E. N. Adams and T.D. Holstein, *J. Phys. Chem. Sol.* **10**, 254 (1959). L.M. Roth and P.N. Argyres, *Semiconductors and semimetals* **1**, 159 (1966) (ed. R.K. Willards and A.C. Beer; New York, Academic Press).
16. P.N. Argyres, *J. Phys. Chem. Sol.* **4**, 19 (1958).
17. F.E. Richards, *Phys. Rev.* **B8**, 2552 (1973).
18. W.M. Becker and H.Y. Fan, *7th International conference on Physics of semiconductors*, p. 663 (Paris, Dunod, 1964).
19. L.S. Lerner, *Phys. Rev.* **130**, 605 (1963).
20. R.J. Sladek, *Phys. Rev.* **110**, 817 (1958).
21. Ref. 1.7, p. 228.
22. R. Kubo, S.J. Miyake and N. Hashitsume, *Solid state Physics*, **17**, 269 (ed. F. Seitz and D. Turnbull; New York, Academic Press, 1965).
23. K. Hiruma, G. Kido and N. Miura, *Sol. State Comm.* **31**, 1019 (1979).
24. S. Tanuma and R. Inada, *Prog. Theor. Phys.* **57**, 231 (1975).
25. L.M. Falicov and R.W. Stark, *Prog. Low. Temp. Phys.* **5**, 235 (1967): surveys the basic ideas and early results.
26. R.G. Chambers, *Proc. Phys. Soc.* **88**, 701 (1966).
27. Ref. 3.12, p. 169. Y. Yafet, *Solid State Physics* **14**, 1 (1963).
28. C. Zener, *Proc. Roy. Soc.* **A145**, 521 (1934).
29. E.I. Blount, *Phys. Rev.* **126**, 1636 (1962).
30. A.B. Pippard, *Proc. Roy. Soc.* **A270**, 1 (1962) and *Phil. Trans. Roy. Soc.* **A265**, 317 (1964). W.G. Chambers, *Phys. Rev.* **140**, A135 (1965).
31. L.M. Falicov, A.B. Pippard and P.R. Sievert, *Phys. Rev.* **151**, 498 (1966).
32. J.B. Ketterson and R.W. Stark, *Phys. Rev.* **156**, 748 (1967).
33. Ref. 1.7, p. 57.
34. J.C. Kimball, R.W. Stark and F.M. Mueller, *Phys. Rev.* **162**, 600 (1967).
35. e.g. C.B. Friedberg, *J.LT.P.* **14**, 147 (1974).
36. E. Brown, *Phys. Rev.* **133**, A1038 (1964). J. Zak, *Phys. Rev.* **134**, A1607 (1964).
37. W.G. Chambers, Ref. 30.
38. D.R. Hofstadter, *Phys. Rev.* **B14**, 2239 (1976).
39. M. Wilkinson, *Proc. Roy. Soc.* **A391**, 305 (1984).
40. A.B. Pippard, *Phil. Trans. Roy. Soc.* **A265**, 317 (1964).
41. L.M. Falicov and H. Stachowiak, *Phys. Rev.* **147**, 505 (1966).
42. J.K. Hulbert and R.C. Young, *Phys. Rev. Lett.* **27**, 1048 (1971).
43. H.W. Capel, *Physica* **42**, 491 (1969); **46**, 169 (1970); **54**, 361 (1971) and **70**, 1 (1973).
44. C.E.T. Gonçalvez da Silva and L.M. Falicov, *Phys. Rev.* **B8**, 527 (1973).
45. J.W. Eddy and R.W. Stark, *Phys. Rev. Lett.* **48**, 275 (1982).
46. A.B. Pippard in *Electrons at the Fermi surface*, p. 124 (ed. M. Springford, Cambridge University Press, 1980).
47. P.L. Taylor, *Phys. Rev.* **B15**, 3558 (1977).
48. R.C. Barklie and A.B. Pippard, *Proc. Roy. Soc.* **A317**, 167 (1970).
49. R.C. Young, *Phys. Rev. Lett.* **27**, 1048 (1971), and *Phys. Lett.* **30A**, 510 (1969).

J.K. Hulbert and R.C. Young, *Low temperature physics* – LT13, **4**, 142 (ed. K.D. Timmerhaus, W.J. O'Sullivan and E.F. Hammel; New York, Plenum, 1974). See also R.C. Young, *Rep. Prog. Phys.* **40**, 1123 (1977).

50. R.C. Young in LT13, p. 146 (see Ref. 49).
51. R.J. Balcombe and R.A. Parker, *Phil. Mag.* **21**, 533 (1970).
52. M.I. Grossbard, Ref. 2.37.
53. F.R. Fickett, Ref. 3.35.
54. R.W. Stark and C.B. Friedberg, *Phys. Rev. Lett.* **26**, 556 (1971); *J.L.T.P.* **14**, 111 and 175 (1974). D. Morrison and R.W. Stark, *J.L.T.P.* **45**, 581 (1981). N.B. Sandesara and R.W. Stark, *Phys. Rev. Lett.* **53**, 1681 (1984).
55. W.A. Reed and J.H. Condon, *Phys. Rev.* **B1**, 3504 (1970).

Chapter 5

1. R. Landauer in *Electrical transport and optical properties of inhomogeneous media*, p. 2 (ed. J.C. Garland and D.B. Tanner; New York, American Institute of Physics, 1978).
   D.S. McLachlan, *J. Phys.* **C20**, 865 (1987).
2. Ref. 1.4.
3. J.M. Ziman, *Phil. Mag.* **3**, 1117 (1958).
4. Ref. 3.30.
5. P.M. Martin, J.B. Sampsell and J.C. Garland, *Phys. Rev.* **B15**, 5598 (1977).
6. J.M. Ziman, *J. Phys.* **C1**, 1532 (1968). See also V.K.S. Shante and S. Kirkpatrick, *Adv. Phys.* **21**, 325 (1971); G.E. Pike and C.H. Seager, *Phys. Rev.* **B10**, 1421 (1974); and J.P. Fitzpatrick, R.B. Malt and F. Spaepen, *Phys. Lett.* **A47**, 207 (1974).
7. Ref. 1.5.
8. Ref. 3.35.
9. Ref. 1.11.
10. A.W. Overhauser, *Can. J. Phys.* **60**, 687 (1982).
11. R. Fletcher, *Can. J. Phys.* **60**, 679 (1982).
12. P.G. Coulter and W.R. Datars, *Can. J. Phys.* **63**, 159 (1985).
13. J. Babiskin and P.G. Siebenmann, *Phys. Kond. Mat.* **9**, 113 (1969).
14. H. Taub, R.L. Schmidt, B.W. Maxfield and R. Bowers, *Phys. Rev.* **B4**, 1134 (1971).
15. R.A. Young, *Phys. Rev.* **175**, 813 (1968).
16. P.M. O'Keefe and W.A. Goddard, *Phys. Rev. Lett.* **23**, 300 (1969).
17. M.R. Stinson, *J. Phys.* **F10**, L133 (1980).
18. Ref. 1.10.
19. R.H. Stokes, *J. Phys. Chem. Sol.* **27**, 51 (1966).
20. G. Stetter, W. Adlhart, G. Fritsch, E. Steichele and E. Lüscher, *J. Phys.* **F8**, 2075 (1978).
21. T.G. Blaney, Ph.D. dissertation, University of Cambridge (1967).
22. Refs. 2.42 and 2.14.
23. J.A. Schaefer and J.A. Marcus, *Phys. Rev. Lett.* **27**, 935 (1971).
24. Ref. 2.43.
25. J.S. Lass, *Phys. Lett.* **39A**, 343 (1972).

References 243

26. F.W. Holroyd and W.R. Datars, *Can. J. Phys.* **53**, 2517 (1975).
27. Ref. 1.7, p. 185.
28. A.W. Overhauser, *Phys. Rev. Lett.* **13**, 190 (1964).
29. Ref. 2.7.
30. I.M. Templeton, *JLTP* **43**, 293 (1981). See Ref. 27 for full discussion.
31. D. Follstaedt and C.P. Slichter, *Phys. Rev.* **B13**, 1017 (1976).
32. W.M. Walsh, L.W. Rupp and P.H. Schmidt, *Phys. Rev.* **142**, 414 (1966).
33. A.W. Overhauser, *Phys. Rev. Lett.* **53**, 64 (1984).
34. J.A. Wilson and M. de Podesta, *J. Phys.* **F16**, L121 (1986).
35. M. Elliott and W.R. Datars, *Sol. State Comm.* **46**, 67 (1983).
36. P.G. Coulter and W.R. Datars, *Sol. State Comm.* **43**, 715 (1982).
37. Ref. 2.35.
38. G.D. Mahan, *J. Phys.* **F14**, 941 (1984).
39. P.T. Coleridge, *J. Phys.* **F17**, L79 (1987).
40. I.M. Templeton, *J. Phys.* **F12**, L121 (1982).

Chapter 6

1. Ref. 1.24, p. 185.
2. T. Ando, A.B. Fowler and F. Stern, *Rev. Mod. Phys.* **54**, 437 (1982) give more than 2000 references. An impression of the subsequent proliferation, especially concerning the quantum Hall effect, can be gained by browsing in the *Proceedings of the 18th international conference on the physics of semiconductors* (ed. O. Engström; Singapore, World Scientific, 1987).
3. E.H. Kennard, *Kinetic theory of gases*, p. 302 (New York, McGraw Hill, 1938).
4. L. Nordheim, *Act. Sci. et Ind.* **131** (1934).
5. R.B. Dingle, *Proc. Roy. Soc.* **A201**, 545 (1950).
6. K. Fuchs, *Proc. Camb. Phil. Soc.* **34**, 100 (1938); E.H. Sondheimer, *Adv. Phys.* **1**, 1 (1952).
7. Ref. 3.35.
8. K. Førsvoll and I. Holwech, *J. Appl Phys.* **34**, 2230 (1963).
9. G. Brandli and J.L. Olsen, *Mat. Sci. Eng.* **4**, 61 (1969).
10. E. Fawcett in Ref. 3.26, p. 197.
11. J.L. Olsen, *Helv. Phys. Acta* **31**, 713 (1958).
12. F.J. Blatt, A. Burmester and B. LaRoy, *Phys. Rev.* **155**, 611 (1967).
13. Ref. 3.6.
14. V.M. Morton, Ph.D. dissertation, Cambridge (1960); D.J. Roaf, *Phil. Trans. Roy. Soc.* **A255**, 135 (1962).
15. A.N. Friedman and S.H. Koenig, *I.B.M. Journ. Res. Dev.* **4**, 158 (1960).
16. A.A. Mitrjaev, O.A. Panchenko, I.I. Razgunov and V.S. Tsoi, *Surf. Sci.* **75**, L376 (1978).
17. K.L. Chopra, *Phys. Rev.* **155**, 660 (1967).
18. H.E. Bennett and J.M. Bennett, in *Optical properties and electronic structure of metals and alloys*, p. 175 (ed. F. Abelès; Amsterdam, North-Holland, 1966).
19. Ref. 1.14.
20. R.G. Chambers, *Proc. Roy. Soc.* **A202**, 378 (1950).
21. Y.-H. Kao, *Phys. Rev.* **138**, A1412 (1965).

22. M. Yaqub and J.F. Cochran, *Phys. Rev. Lett.* **10**, 390 (1963), and *Phys. Rev.* **137**, A1182 (1965).

23. O.S. Lutes and D.A. Clayton, *Phys. Rev.* **138**, A1448 (1965).

24. M.Ya. Azbel' and V.G. Peschanskii, *J.E.T.P.* **22**, 399 (1966).

25. E. Ditlefsen and J. Lothe, *Phil. Mag.* **14**, 759 (1966).

26. Ref. 1.24, pp. 220, 233.

27. Ref. 24 and references in V.G. Peschanskii and M.Ya. Azbel', *J.E.T.P.* **28**, 1045 (1969).

28. K. Førsvoll and I. Holwech, *Phil. Mag.* **9**, 435 (1964).

29. H.J. Mackey, J.R. Sybert and W.D. Deering, *Phys. Rev.* **176**, 857 (1968).

30. E.H. Sondheimer, *Phys. Rev.* **80**, 401 (1950), and Ref. 6.

31. V.L. Gurevich, *J.E.T.P.* **8**, 464 (1958).

32. Ref. 3.9.

33. J. Babiskin and P.G. Siebenmann, *Phys. Rev.* **107**, 1249 (1957).

34. J.A. Munarin, J.A. Marcus and P.E. Bloomfield, *Phys. Rev.* **172**, 718 (1968).

35. P.D. Hambourger, J.A. Marcus and J.A. Munarin, *Phys. Lett.* **25A**, 461 (1967).

36. J.M. Reynolds, K.R. Efferson, C.G. Grenier and N.H. Zebouni, in Ref. 1.15, p. 808; C.G. Grenier, K.R. Efferson and J.M. Reynolds, *Phys. Rev.* **143**, 406 (1966).

37. A.B. Pippard, *Phil. Mag.* **13**, 1143 (1966).

38. Yu. V. Sharvin, *J.E.T.P.* **21**, 655 (1965).

39. R.A.R. Tricker, *Proc. Camb. Phil. Soc.* **22**, 454 (1925).

40. Yu.V. Sharvin and N.I. Bogatina, *J.E.T.P.* **29**, 419 (1969).

41. V.S. Tsoi, *J.E.T.P. Lett.* **19**, 70 (1974).

42. A.B. Pippard, *Proc. Roy. Soc.* **A191**, 385 (1947).

43. E. Jahnke and F. Emde, *Tables of functions*, p. 251 (6th edition, ed. F. Lösch) (New York, McGraw-Hill, 1960).

44. V.S. Tsoi and I.I. Razgunov, *J.E.T.P. Lett.* **25**, 26 (1977).

45. M.K. Debe and D.A. King, Surf. Sci. **81**, 193 (1979).

46. F.M. Hawkins (with appendix by A.B. Pippard), *Proc. Camb. Phil. Soc.* **61**, 433 (1965).

47. J. Clarke and L.A. Schwartzkopf, *J.L.T.P.* **16**, 317 (1974).

48. Y. Imry, in *Directions in condensed matter physics*, p. 101 (ed. G. Grinstein and G. Mazenko) (Singapore, World Scientific, 1986).

49. C.P. Umbach, S. Washburn, R.B. Laibowitz and R.A. Webb, *Phys. Rev.* **B30**, 4048 (1984).

50. P.W. Anderson, *Phil. Mag.* **52**, 505 (1985).

51. A.B. Pippard, *The physics of vibration*, vol. 1, p. 81 (Cambridge University Press, 1978).

52. Ref. 51, p. 88.

53. Y. Aharonov and D. Bohm, *PHys. Rev.* **115**, 485 (1959).

54. R. Doll and M. Näbauer, *Phys. Rev. Lett.* **7**, 51 (1961).

55. R. Landauer and M. Büttiker, *Phys. Rev.* **B36**, 6255 (1987).

56. Ref. 51, p. 92.

57. A.D. Stone, *Phys. Rev. Lett.* **54**, 2692 (1985); Y. Imry, *Europhys. Lett.* **1**, 249 (1986), and Ref. 48.

58. R.A. Webb, S. Washburn, C.P. Umbach and R.B. Laibowitz, *Phys. Rev. Lett.* **54**, 2696 (1985).
59. V. Chandrasekhar, M.J. Rooks, S. Wind and D.E. Prober, *Phys. Rev. Lett.* **55**, 1610 (1985).
60. B.L. Al'tshuler, A.G. Aronov and B.Z. Spivak, *J.E.T.P. Lett.* **33**, 94 (1981).
61. D.Yu. Sharvin and Yu.V. Sharvin, *J.E.T.P. Lett.* **34**, 272 (1981).

Index of names

(Chapter reference numbers indicated in brackets, followed by page references)

248   *Index of names*

Subject index

Definitions

In addition to the definitions and symbols tabulated on p. xi, the following are to be found in the text.

Materials